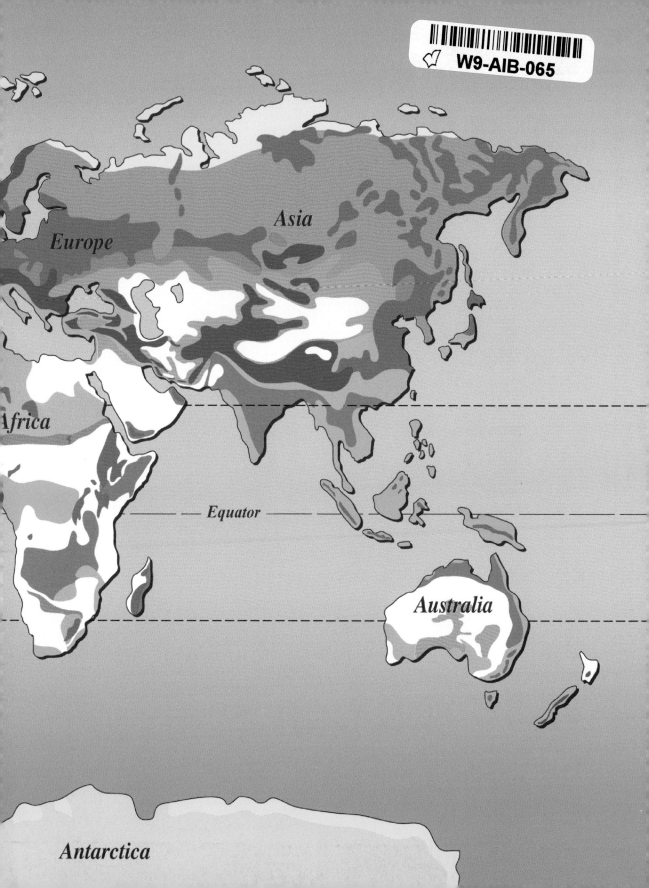

U·X·L
ENCYCLOPEDIA OF
BIOMES

U·X·L
ENCYCLOPEDIA OF
BIOMES

MARLENE WEIGEL

3

**RIVER AND STREAM
SEASHORE
TUNDRA
WETLAND**

AN IMPRINT OF THE GALE GROUP

DETROIT · SAN FRANCISCO · LONDON
BOSTON · WOODBRIDGE, CT

U•X•L Encyclopedia of Biomes
Marlene Weigel

STAFF

Julie L. Carnagie, *U•X•L Editor*
Carol DeKane Nagel, *U•X•L Managing Editor*
Thomas L. Romig, *U•X•L Publisher*
Meggin Condino, *Senior Analyst, New Product Development*

Margaret Chamberlain, *Permissions Specialist* (Pictures)

Rita Wimberley, *Senior Buyer*
Evi Seoud, *Assistant Production Manager*
Dorothy Maki, *Manufacturing Manager*

Michelle DiMercurio, *Art Director*
Cynthia Baldwin, *Product Design Manager*

Graphix Group, *Typesetting*

Library of Congress Cataloging-in-Publication Data

U•X•L encyclopedia of biomes / Marlene Weigel, Julie L. Carnagie, editor.
    p.   cm.
    Includes bibliographical references and indexes.
    Contents: v. 1. Coniferous forests, continental margins, deciduous forests, and deserts –
    v. 2. Grasslands, lakes and ponds, oceans, and rainforests – v. 3. Rivers, seashores, tundras, and wetlands.
    ISBN 0-7876-3732-7 (set). -- ISBN 0-7876-3733-5 (vol. 1). – ISBN 0-7876-3734-3
    (vol. 2). – ISBN 0-7876-3735-1 (vol. 3)
    1. Biotic communities Juvenile literature. [1. Biotic communities.] I. Weigel, Marlene. II.
    Carnagie, Julie. III. Title: Encyclopedia of biomes.
    QH541.14.U18 1999
    577.8'2–DC21
99-23395   CIP

Printed in the United States of America
10 9 8 7 6 5 4 3 2

TABLE OF CONTENTS

READER'S GUIDE

U•X•L Encyclopedia of Biomes offers readers comprehensive and easy-to-use information on twelve of the Earth's major biomes and their many component ecosystems. Arranged alphabetically across three volumes, each biome chapter includes an overview; a description of how the biomes are formed; their climate; elevation; growing season; plants, animals, and endangered species; food webs; human culture; and economy. The information presented may be used in a variety of subject areas, such as biology, geography, anthropology, and current events. Each chapter includes a "spotlight" feature focusing on specific geographical areas related to the biome being discussed and concludes with a section composed of books, periodicals, Internet addresses, and environmental organizations for readers to conduct more extensive research.

ADDITIONAL FEATURES

Each volume of *U•X•L Encyclopedia of Biomes* includes a color-coded world biome map and a sixteen-page color insert. More than120 black-and-white photos, illustrations, and maps enliven the text, while sidebar boxes highlight fascinating facts and related information. All three volumes include a glossary; a bibliography; and a subject index covering all the subjects discussed in *U•X•L Encyclopedia of Biomes*.

NOTE

There are many different ways to describe certain aspects of a particular biome, and it would be impossible to include all of the classifications in *U•X•L Encyclopedia of Biomes*. However, in cases where more than one classification seemed useful, more than one was given. Please note that the classifications represented here may not be those preferred by all specialists in a particular field.

Every effort was made in this set to include the most accurate information pertaining to sizes and other measurements. Great variations exist in the available data, however. Sometimes differences can be accounted for in terms of what was measured: the reported area of a lake, for example, may vary depending upon the point at which measuring began. Other differences may result from natural changes that took place between the time of one measurement and another. Further, other data may be questionable because reliable information has been difficult to obtain. This is particularly true for remote areas in developing countries, where funds for scientific research are lacking and non-native scientists may not be welcomed.

ACKNOWLEDGMENTS

Special thanks are due for the invaluable comments and suggestions provided by the *U•X•L Encyclopedia of Biomes* advisors:

- Nancy Bard, Librarian, Thomas Jefferson High School for Science and Technology, Alexandria, Virginia

- Frances L. Cohen, Consultant, Malvern, Pennsylvania

- Valerie Doud, Science Teacher, Peru Junior High School, Peru, Indiana

- Elaine Ezell, Library Media Specialist, Bowling Green Junior High School, Bowling Green, Ohio

The author and editor would like to thank contributing writer Rita Travis for her work on the Coniferous Forest, Grassland, Tundra, and Wetland chapters.

COMMENTS AND SUGGESTIONS

We welcome your comments on this work as well as your suggestions for topics to be featured in future editions of *U•X•L Encyclopedia of Biomes.* Please write: Editors, *U•X•L Encyclopedia of Biomes,* U•X•L, 27500 Drake Rd., Farmington Hills, MI 48331-3535; call toll-free: 1-800-877-4253; fax: 248-699-8097; or send e-mail via www.galegroup.com.

WORDS TO KNOW

A

Abyssal plain: The flat midportion of the ocean floor that begins beyond the continental rise.

Acid rain: A mixture of water vapor and polluting compounds in the atmosphere that falls to the Earth as rain or snow.

Active margin: A continental margin constantly being changed by earthquake and volcanic action.

Aerial roots: Plant roots that dangle in midair and absorb nutrients from their surroundings rather than from the soil.

Algae: Plantlike organisms that usually live in watery environments and depend upon photosynthesis for food.

Algal blooms: Sudden increases in the growth of algae on the ocean's surface.

Alluvial fan: A fan-shaped area created when a river or stream flows downhill, depositing sediment into a broader base that spreads outward.

Amphibians: Animals that spend part, if not most, of their lives in water.

Amphibious: Able to live on land or in water.

Angiosperms: Trees that bear flowers and produce their seeds inside a fruit; deciduous and rain forest trees are usually angiosperms.

Annuals: Plants that live for only one year or one growing season.

Aquatic: Having to do with water.

Aquifier: Rock beneath the Earth's surface in which groundwater is stored.

Arachnids: Class of animals that includes spiders and scorpions.

Arctic tundra: Tundra located in the far north, close to or above the Arctic Circle.

Arid: Dry.

Arroyo: The dry bed of a stream that flows only after rain; also called a wash or a *wadi*.

Artifacts: Objects made by humans, including tools, weapons, jars, and clothing.

Artificial grassland: A grassland created by humans.

Artificial wetland: A wetland created by humans.

Atlantic blanket bogs: Blanket bogs in Ireland that are less than 656 feet (200 meters) above sea level.

Atolls: Ring-shaped reefs formed around a lagoon by tiny animals called corals.

B

Bactrian camel: The two-humped camel native to central Asia.

Bar: An underwater ridge of sand or gravel formed by tides or currents that extends across the mouth of a bay.

Barchan dunes: Sand dunes formed into crescent shapes with pointed ends created by wind blowing in the direction of their points.

Barrier island: An offshore island running parallel to a coastline that helps shelter the coast from the force of ocean waves.

Barrier reef: A type of reef that lines the edge of a continental shelf and separates it from deep ocean water. A barrier reef may enclose a lagoon and even small islands.

Bathypelagic zone: An oceanic zone based on depth that ranges from 3,300 to 13,000 feet (1,000 to 4,000 meters).

Bathyscaphe: A small, manned, submersible vehicle that accommodates several people and is able to withstand the extreme pressures of the deep ocean.

Bay: An area of the ocean partly enclosed by land; its opening into the ocean is called a mouth.

Beach: An almost level stretch of land along a shoreline.

Bed: The bottom of a river or stream channel.

Benthic: Term used to describe plants or animals that live attached to the seafloor.

Biodiverse: Term used to describe an environment that supports a wide variety of plants and animals.

Bio-indicators: Plants or animals whose health is used to indicate the general health of their environment.

Biological productivity: The growth rate of life forms in a certain period of time.

Biome: A distinct, natural community chiefly distinguished by its plant life and climate.

Blanket bogs: Shallow bogs that spread out like a blanket; they form in areas with relatively high levels of annual rainfall.

Bog: A type of wetland that has wet, spongy, acidic soil called peat.

Boreal forest: A type of coniferous forest found in areas bordering the Arctic tundra. Also called taiga.

Boundary layer: A thin layer of water along the floor of a river channel where friction has stopped the flow completely.

Brackish water: A mixture of freshwater and saltwater.

Braided stream: A stream consisting of a network of interconnecting channels broken by islands or ridges of sediment, primarily mud, sand, or gravel.

Branching network: A network of streams and smaller rivers that feeds a large river.

Breaker: A wave that collapses on a shoreline because the water at the bottom is slowed by friction as it travels along the ocean floor and the top outruns it.

Browsers: Herbivorous animals that eat from trees and shrubs.

Buoyancy: Ability to float.

Buran: Strong, northeasterly wind that blows over the Russian steppes.

Buttresses: Winglike thickenings of the lower trunk that give tall trees extra support.

C

Canopy: A roof over the forest created by the foliage of the tallest trees.

Canyon: A long, narrow valley between high cliffs that has been formed by the eroding force of a river.

Carbon cycle: Natural cycle in which trees remove excess carbon dioxide from the air and use it during photosynthesis. Carbon is then returned to the soil when trees die and decay.

Carnivore: A meat-eating plant or animal.

Carrion: Decaying flesh of dead animals.

Cay: An island formed from a coral reef.

Channel: The path along which a river or stream flows.

Chemosynthesis: A chemical process by which deep-sea bacteria use organic compounds to obtain food and oxygen.

Chernozim: A type of temperate grassland soil; also called black earth.

Chinook: A warm, dry wind that blows over the Rocky Mountains in North America.

Chitin: A hard chemical substance that forms the outer shell of certain invertebrates.

Chlorophyll: The green pigment in leaves used by plants to turn energy from the Sun into food.

Clear-cutting: The cutting down of every tree in a selected area.

Climax forest: A forest in which only one species of tree grows because it has taken over and only that species can survive there.

Climbers: Plants that have roots in the ground but use hooklike tendrils to climb on the trunks and limbs of trees in order to reach the canopy, where there is light.

Cloud forest: A type of rain forest that occurs at elevations over 10,500 feet (3,200 meters) and that is covered by clouds most of the time.

Commensalism: Relationship between organisms in which one reaps a benefit from the other without either harming or helping the other.

Commercial fishing: Fishing done to earn money.

Coniferous trees: Trees, such as pines, spruces, and firs, that produce seeds within a cone.

Consumers: Animals in the food web that eat either plants or other animals.

Continental shelf: A flat extension of a continent that tapers gently into the sea.

Continental slope: An extension of a continent beyond the continental shelf that dips steeply into the sea.

Convergent evolution: When distantly related animals in different parts of the world evolve similar characteristics.

Coral reef: A wall formed by the skeletons of tiny animals called corals.

Coriolis Effect: An effect on wind and current direction caused by the Earth's rotation.

Crustaceans: Invertebrate animals that have hard outer shells.

Current: The steady flow of water in a certain direction.

D

Dambo: Small marsh found in Africa.

Dark zone: The deepest part of the ocean, where no light reaches.

Deciduous: Term used to describe trees, such as oaks and elms, that lose their leaves during cold or very dry seasons.

Decompose: The breaking down of dead plants and animals in order to release nutrients back into the environment.

Decomposers: Organisms that feed on dead organic materials, releasing nutrients into the environment.

Dehydration: Excessive loss of water from the body.

Delta: Muddy sediments that have formed a triangular shape over the continental shelf near the mouth of a river.

Deposition: The carrying of sediments by a river from one place to another and depositing them.

Desalination: Removing the salt from seawater.

Desert: A very dry area receiving no more than 10 inches (25 centimeters) of rain during a year and supporting little plant or animal life.

Desertification: The changing of fertile lands into deserts through destruction of vegetation (plant life) or depletion of soil nutrients. Topsoil and groundwater are eventually lost as well.

Desert varnish: A dark sheen on rocks and sand believed to be caused by the chemical reaction between overnight dew and minerals in the soil.

Diatom: A type of phytoplankton with a geometric shape and a hard, glasslike shell.

Dinoflagellate: A type of phytoplankton having two whiplike attachments that whirl in the water.

Discharge: The amount of water that flows out of a river or stream into another river, a lake, or the ocean.

Doldrums: Very light winds near the equator that create little or no movement in the ocean.

Downstream: The direction toward which a river or stream is flowing.

Drainage basin: All the land area that supplies water to a river or stream.

Dromedary: The one-humped, or Arabian, camel.

Drought: A long, extremely dry period.

Dune: A hill or ridge of sand created by the wind.

Duricrusts: Hard, rocklike crusts on ridges that are formed by a chemical reaction caused by the combination of dew and minerals such as limestone.

E

Ecosystem: A network of organisms that have adapted to a particular environment.

Eddy: A current that moves against the regular current, usually in a circular motion.

Elevation: The height of an object in relation to sea level.

Emergent plants: Plants that are rooted at the bottom of a body of water that have top portions that appear to be above the water's surface.

Emergents: The very tallest trees in the rain forest, which tower above the canopy.

Engineered wood: Manufactured wood products composed of particles of several types of wood mixed with strong glues and preservatives.

Epilimnion: The layer of warm or cold water closest to the surface of a large lake.

Epipelagic zone: An oceanic zone based on depth that reaches down to 650 feet (200 meters).

Epiphytes: Plants that grow on other plants or hang on them for physical support.

Ergs: Arabian word for vast seas of sand dunes, especially those found in the Sahara.

Erosion: Wearing away of the land.

Estivation: An inactive period experienced by some animals during very hot months.

Estuary: The place where a river traveling through lowlands meets the ocean in a semi-enclosed area.

Euphotic zone: The zone in a lake where sunlight can reach.

Eutrophication: Loss of oxygen in a lake or pond because increased plant growth has blocked sunlight.

F

Fast ice: Ice formed on the surface of the ocean between pack ice and land.

Faults: Breaks in the Earth's crust caused by earthquake action.

Fell-fields: Bare rock-covered ground in the alpine tundra.

Fen: A bog that lies at or below sea level and is fed by mineral-rich groundwater.

First-generation stream: The type of stream on which a branching network is based; a stream with few tributaries. Two first-generation streams join to form a second-generation stream and so on.

Fish farms: Farms in which fish are raised for commercial use; also called hatcheries.

Fjords: Long, narrow, deep arms of the ocean that project inland.

Flash flood: A flood caused when a sudden rainstorm fills a dry riverbed to overflowing.

Floating aquatic plant: A plant that floats either partly or completely on top of the water.

Flood: An overflow caused when more water enters a river or stream than its channel can hold at one time.

Floodplain: Low-lying, flat land easily flooded because it is located next to streams and rivers.

Food chain: The transfer of energy from organism to organism when one organism eats another.

Food web: All of the possible feeding relationships that exist in a biome.

Forbs: A category of flowering, broadleaved plants other than grasses that lack woody stems.

Forest: A large number of trees covering not less than 25 percent of the area where the tops of the trees interlock, forming a canopy at maturity.

Fossil fuels: Fuels made from oil and gas that formed over time from sediments made of dead plants and animals.

Fossils: Remains of ancient plants or animals that have turned to stone.

Freshwater lake: A lake that contains relatively pure water and relatively little salt or soda.

Freshwater marsh: A wetland fed by freshwater and characterized by poorly drained soil and plant life dominated by nonwoody plants.

Freshwater swamp: A wetland fed by freshwater and characterized by poorly drained soil and plant life dominated by trees.

Friction: The resistance to motion when one object rubs against another.

Fringing reef: A type of coral reef that develops close to the land; no lagoon separates it from the shore.

Frond: A leaflike organ found on all species of kelp plants.

Fungi: Plantlike organisms that cannot make their own food by means of photosynthesis; instead they grow on decaying organic matter or live as parasites on a host.

G

Geyser: A spring heated by volcanic action. Some geysers produce enough steam to cause periodic eruptions of water.

Glacial moraine: A pile of rocks and sediments created as a glacier moves across an area.

Global warming: Warming of the Earth's climate that may be speeded up by air pollution.

Gorge: A deep, narrow pass between mountains.

Grassland: A biome in which the dominant vegetation is grasses rather than trees or tall shrubs.

Grazers: Herbivorous animals that eat low-growing plants such as grass.

Ground birds: Birds that hunt food and make nests on the ground or close to it.

Groundwater: Freshwater stored in rock layers beneath the ground.

Gulf: A large area of the ocean partly enclosed by land; its opening is called a strait.

Gymnosperms: Trees that produce seeds that are often collected together into cones; most conifers are gymnosperms.

Gyre: A circular or spiral motion.

H

Hadal zone: An oceanic zone based on depth that reaches from 20,000 to 35,630 feet (6,000 to 10,860 meters).

Hardwoods: Woods usually produced by deciduous trees, such as oaks and elms.

Hatcheries: Farms in which fish are raised for commercial use; also called fish farms.

Headland: An arm of land made from hard rock that juts out into the ocean after softer rock has been eroded away by the force of tides and waves.

Headwaters: The source of a river or stream.

Herbicides: Poisons used to control weeds or any other unwanted plants.

Herbivore: An animal that eats only plant matter.

Herders: People who raise herds of animals for food and other needs; they may also raise some crops but are usually not dependent upon them.

Hermaphroditic: Term used to describe an animal or plant in which reproductive organs of both sexes are present in one individual.

Hibernation: An inactive period experienced by some animals during very cold months.

High tide: A rising of the surface level of the ocean caused by the Earth's rotation and the gravitational pull of the Sun and Moon.

Holdfast: A rootlike structure by which kelp plants anchor themselves to rocks or the seafloor.

Hummocks: 1. Rounded hills or ridges, often heavily wooded, that are higher than the surrounding area; 2. Irregularly shaped ridges formed when large blocks of ice hit each other and one slides on top of the other; also called hammocks.

Humus: The nutrient-rich, spongy matter produced when the remains of plants and animals are broken down to form soil.

Hunter-gatherers: People who live by hunting animals and gathering nuts, berries, and fruits; normally, they do not raise crops or animals.

Hurricane: A violent tropical storm that begins over the ocean.

Hydric soil: Soil that contains a lot of water but little oxygen.

Hydrologic cycle: The manner in which molecules of water evaporate, condense as clouds, and return to the Earth as precipitation.

Hydrophytes: Plants that are adapted to grow in water or very wet soil.

Hypolimnion: The layer of warm or cold water closest to the bottom of a large lake.

Hypothermia: A lowering of the body temperature that can result in death.

I

Insecticides: Poisons that kill insects.

Insectivores: Plants and animals that feed on insects.

Intermittent stream: A stream that flows only during certain seasons.

Interrupted stream: A stream that flows aboveground in some places and belowground in others.

Intertidal zone: The seashore zone covered with water during high tide and dry during low tide; also called the middle, or the littoral, zone.

Invertebrates: Animals without a backbone.

K

Kelp: A type of brown algae that usually grows on rocks in temperate water.

Kettle: A large pit created by a glacier that fills with water and becomes a pond or lake.

Kopjes: Small hills made out of rocks that are found on African grasslands.

L

Labrador Current: An icy Arctic current that mixes with warmer waters off the coast of northeastern Canada.

Lagoon: A large pool of seawater cut off from the ocean by a bar or other landmass.

Lake: A usually large body of inland water that is deep enough to have two distinct layers based on temperature.

Latitude: A measurement on a map or globe of a location north or south of the equator. The measurements are made in degrees, with the equator, or dissecting line, being zero.

Layering: Tree reproduction that occurs when a branch close to the ground develops roots from which a new tree grows.

Levee: High bank of sediment deposited by a very silty river.

Lichens: Plantlike organisms that are combinations of algae and fungi. The algae produces the food for both by means of photosynthesis.

Limnetic zone: The deeper, central region of a lake or pond where no plants grow.

Littoral zone: The area along the shoreline that is exposed to the air during low tide; also called intertidal zone.

Longshore currents: Currents that move along a shoreline.

Lowland rain forest: Rain forest found at elevations up to 3,000 feet (900 meters).

Low tide: A lowering of the surface level of the ocean caused by the Earth's rotation and the gravitational pull of the Sun and Moon.

M

Macrophytic: Term used to describe a large plant.

Magma: Molten rock from beneath the Earth's crust.

Mangrove swamp: A coastal saltwater swamp found in tropical and subtropical areas.

Marine: Having to do with the oceans.

Marsh: A wetland characterized by poorly drained soil and by plant life dominated by nonwoody plants.

Mature stream: A stream with a moderately wide channel and sloping banks.

Meandering stream: A stream that winds snakelike through flat countryside.

Mesopelagic zone: An oceanic zone based on depth that ranges from 650 to 3,300 feet (200 to 1,000 meters).

Mesophytes: Plants that live in soil that is moist but not saturated.

Mesophytic: Term used to describe a forest that grows where only a moderate amount of water is available.

Mid-ocean ridge: A long chain of mountains that lies under the World Ocean.

Migratory: Term used to describe animals that move regularly from one place to another in search of food or to breed.

Mixed-grass prairie: North American grassland with a variety of grass species of medium height.

Montane rain forest: Mountain rain forest found at elevations between 3,000 and 10,500 feet (900 and 3,200 meters).

Mountain blanket bogs: Blanket bogs in Ireland that are more than 656 feet (200 meters) above sea level.

Mouth: The point at which a river or stream empties into another river, a lake, or an ocean.

Muck: A type of gluelike bog soil formed when fully decomposed plants and animals mix with wet sediments.

Muskeg: A type of wetland containing thick layers of decaying plant matter.

Mycorrhiza: A type of fungi that surrounds the roots of conifers, helping them absorb nutrients from the soil.

N

Neap tides: High tides that are lower and low tides that are higher than normal when the Earth, Sun, and Moon form a right angle.

Nekton: Animals that can move through the water without the help of currents or wave action.

Neritic zone: That portion of the ocean that lies over the continental shelves.

Nomads: People or animals who have no permanent home but travel within a well-defined territory determined by the season or food supply.

North Atlantic Drift: A warm ocean current off the coast of northern Scandinavia.

Nutrient cycle: Natural cycle in which mineral nutrients are absorbed from the soil by tree roots and returned to the soil when the tree dies and the roots decay.

O

Oasis: A fertile area in the desert having a water supply that enables trees and other plants to grow there.

Ocean: The large body (or bodies) of saltwater that covers more than 70 percent of the Earth's surface.

Oceanography: The exploration and scientific study of the oceans.

Old stream: A stream with a very wide channel and banks that are nearly flat.

Omnivore: Organism that eats both plants and animals.

Ooze: Sediment formed from the dead tissues and waste products of marine plants and animals.

Oxbow lake: A curved lake formed when a river abandons one of its bends.

Oxygen cycle: Natural cycle in which the oxygen taken from the air by plants and animals is returned to the air by plants during photosynthesis.

P

Pack ice: A mass of large pieces of floating ice that have come together on an open ocean.

Pamir: A high altitude grassland in Central Asia.

Pampa: A tropical grassland found in South America.

Pampero: A strong, cold wind that blows down from the Andes Mountains and across the South American pampa.

Pantanal: A wet savanna that runs along the Upper Paraguay River in Brazil.

Parasite: An organism that depends upon another organism for its food or other needs.

Passive margin: A continental margin free of earthquake or volcanic action, and in which few changes take place.

Peat: A type of soil formed from slightly decomposed plants and animals.

Peatland: Wetlands characterized by a type of soil called peat.

Pelagic zone: The water column of the ocean.

Perennials: Plants that live at least two years or seasons, often appearing to die but returning to "life" when conditions improve.

Permafrost: Permanently frozen topsoil found in northern regions.

Permanent stream: A stream that flows continually, even during a long, dry season.

Pesticides: Poisons used to kill anything that is unwanted and is considered a pest.

Phosphate: An organic compound used in making fertilizers, chemicals, and other commercial products.

Photosynthesis: The process by which plants use the energy from sunlight to change water (from the soil) and carbon dioxide (from the air) into the sugars and starches they use for food.

Phytoplankton: Tiny, one-celled algae that float on ocean currents.

Pingos: Small hills formed when groundwater freezes.

Pioneer trees: The first trees to appear during primary succession; they include birch, pine, poplar, and aspen.

Plankton: Plants or animals that float freely in ocean water; from the Greek word meaning "wanderer."

Plunge pool: Pool created where a waterfall has gouged out a deep basin.

Pocosin: An upland swamp whose only source of water is rain.

Polar climate: A climate with an average temperature of not more than 50°F (10°C) in July.

Polar easterlies: Winds occurring near the Earth's poles that blow in an easterly direction.

Pollination: The carrying of pollen from the male reproductive part of a plant to the female reproductive part of a plant so that reproduction may occur.

Polygons: Cracks formed when the ground freezes and contracts.

Pond: A body of inland water that is usually small and shallow and has a uniform temperature throughout.

Pool: A deep, still section in a river or stream.

Prairie: A North American grassland that is defined by the stature of grass groups contained within (tall, medium, and short).

Prairie potholes: Small marshes no more than a few feet deep.

Precipitation: Rain, sleet, or snow.

Predatory birds: Carnivores that hunt food by soaring high in the sky to obtain a "bird's-eye" view and then swoop down to capture their prey.

Primary succession: Period of plant growth that begins when nothing covers the land except bare sand or soil.

Producers: Plants and other organisms in the food web of a biome that are able to make food from nonliving materials, such as the energy from sunlight.

Profundal zone: The zone in a lake where no more than 1 percent of sunlight can penetrate.

Puna: A high-altitude grassland in the Andes Mountains of South America.

R

Rain forest: A tropical forest, or jungle, with a warm, wet climate that supports year-round tree growth.

Raised bogs: Bogs that grow upward and are higher than the surrounding area.

Rapids: Fast-moving water created when softer rock has been eroded to create many short drops in the channel; also called white water.

Reef: A ridge or wall of rock or coral lying close to the surface of the ocean just off shore.

Rhizomes: Plant stems that spread out underground and grow into a new plant that breaks above the surface of the soil or water.

Rice paddy: A flooded field in which rice is grown.

Riffle: A stretch of rapid, shallow, or choppy water usually caused by an obstruction, such as a large rock.

Rill: Tiny gully caused by flowing water.

Riparian marsh: Marsh usually found along rivers and streams.

Rip currents: Strong, dangerous currents caused when normal currents moving toward shore are deflected away from it through a narrow channel; also called riptides.

River: A natural flow of running water that follows a well-defined, permanent path, usually within a valley.

River system: A river and all its tributaries.

S

Salinity: The measure of salts in ocean water.

Salt lake: A lake that contains more than 0.1 ounce of salt per quart (3 grams per liter) of water.

Salt pan: The crust of salt left behind when a salt or soda lake or pond dries up.

Saltwater marsh: A wetland fed by saltwater and characterized by poorly drained soil and plant life dominated by nonwoody plants.

Saltwater swamp: A wetland fed by saltwater and characterized by poorly drained soil and plant life dominated by trees.

Saturated: Soaked with water.

Savanna: A grassland found in tropical or subtropical areas, having scattered trees and seasonal rains.

Scavenger: An animal that eats decaying matter.

School: Large gathering of fish.

Sea: A body of saltwater smaller and shallower than an ocean but connected to it by means of a channel; sea is often used interchangeably with ocean.

Seafloor: The ocean basins; the area covered by ocean water.

Sea level: The height of the surface of the sea. It is used as a standard in measuring the heights and depths of other locations such as mountains and oceans.

Seamounts: Isolated volcanoes on the ocean floor that do not break the surface of the ocean.

Seashore: The strip of land along the edge of an ocean.

Secondary succession: Period of plant growth occurring after the land has been stripped of trees.

Sediments: Small, solid particles of rock, minerals, or decaying matter carried by wind or water.

Seiche: A wave that forms during an earthquake or when a persistent wind pushes the water toward the downwind end of a lake.

Seif dunes: Sand dunes that form ridges lying parallel to the wind; also called longitudinal dunes.

Shelf reef: A type of coral reef that forms on a continental shelf having a hard, rocky bottom. A shallow body of water called a lagoon may be located between the reef and the shore.

Shoals: Areas where enough sediments have accumulated in the river channel that the water is very shallow and dangerous for navigation.

Shortgrass prairie: North American grassland on which short grasses grow.

Smokers: Jets of hot water expelled from clefts in volcanic rock in the deep-seafloor.

Soda lake: A lake that contains more than 0.1 ounce of soda per quart (3 grams per liter) of water.

Softwoods: Woods usually produced by coniferous trees.

Sonar: The use of sound waves to detect objects.

Soredia: Algae cells with a few strands of fungus around them.

Source: The origin of a stream or river.

Spit: A long narrow point of deposited sand, mud, or gravel that extends into the water.

Spores: Single plant cells that have the ability to grow into a new organism.

Sport fishing: Fishing done for recreation.

Spring tides: High tides that are higher and low tides that are lower than normal because the Earth, Sun, and Moon are in line with one another.

Stagnant: Term used to describe water that is unmoving and contains little oxygen.

Star-shaped dunes: Dunes created when the wind comes from many directions; also called stellar dunes.

Steppe: A temperate grassland found mostly in southeast Europe and Asia.

Stone circles: Piles of rocks moved into a circular pattern by the expansion of freezing water.

Straight stream: A stream that flows in a straight line.

Strait: The shallow, narrow channel that connects a smaller body of water to an ocean.

Stream: A natural flow of running water that follows a temporary path that is not necessarily within a valley; also called brook or creek. Scientists often use the term to mean any natural flow of water, including rivers.

Subalpine forest: Mountain forest that begins below the snow line.

Subduction zone: Area where pressure forces the seafloor down and under the continental margin, often causing the formation of a deep ocean trench.

Sublittoral zone: The seashore's lower zone, which is underwater at all times, even during low tide.

Submergent plant: A plant that grows entirely beneath the water.

Subpolar gyre: The system of currents resulting from winds occurring near the poles of the Earth.

Subsistence fishing: Fishing done to obtain food for a family or a community.

Subtropical: Term used to describe areas bordering the equator in which the weather is usually warm.

Subtropical gyre: The system of currents resulting from winds occurring in subtropical areas.

Succession: The process by which one type of plant or tree is gradually replaced by others.

Succulents: Plants that appear thick and fleshy because of stored water.

Sunlit zone: The uppermost part of the ocean that is exposed to light; it reaches down to about 650 feet (200 meters) deep.

Supralittoral zone: The seashore's upper zone, which is never underwater, although it may be frequently sprayed by breaking waves; also called the splash zone.

Swamp: A wetland characterized by poorly drained soil, stagnant water, and plant life dominated by trees.

Sward: Fine grasses that cover the soil.

Swell: Surface waves that have traveled for long distances and become more regular in appearance and direction.

T

Taiga: Coniferous forest found in areas bordering the Arctic tundra; also called boreal forest.

Tallgrass prairie: North American grassland on which only tall grass species grow.

Tannins: Chemical substances found in the bark, roots, seeds, and leaves of many plants and used to soften leather.

Tectonic action: Movement of the Earth's crust, as during an earthquake.

Temperate bog: Peatland found in temperate climates.

Temperate climate: Climate in which summers are hot and winters are cold, but temperatures are seldom extreme.

Temperate zone: Areas in which summers are hot and winters are cold but temperatures are seldom extreme.

Thermal pollution: Pollution created when heated water is dumped into the ocean. As a result, animals and plants that require cool water are killed.

Thermocline: Area of the ocean's water column, beginning at about 1,000 feet (300 meters), in which the temperature changes very slowly.

Thermokarst: Shallow lakes in the Arctic tundra formed by melting permafrost; also called thaw lakes.

Tidal bore: A surge of ocean water caused when ridges of sand direct the ocean's flow into a narrow river channel, sometimes as a single wave.

Tidepools: Pools of water that form on a rocky shoreline during high tide and that remain after the tide has receded.

Tides: Rhythmic movements, caused by the Earth's rotation and the gravitational pull of the Sun and Moon, that raise or lower the surface level of the oceans.

Tombolo: A bar of sand that has formed between the beach and an island, linking them together.

Trade winds: Winds occurring both north and south of the equator to about 30 degrees latitude; they blow primarily east.

Transverse dunes: Sand dunes lying at right angles to the direction of the wind.

Tree: A large woody perennial plant with a single stem, or trunk, and many branches.

Tree line: The elevation above which trees cannot grow.

Tributary: A river or stream that flows into another river or stream.

Tropical: Term used to describe areas close to the equator in which the weather is always warm.

Tropical tree bog: Bog found in tropical climates in which peat is formed from decaying trees.

Tropic of Cancer: A line of latitude about 25 degrees north of the equator.

Tropic of Capricorn: A line of latitude about 25 degrees south of the equator.

Tsunami: A huge wave or upwelling of water caused by undersea earthquakes that grows to great heights as it approaches shore.

Tundra: A cold, dry, windy region where trees cannot grow.

Turbidity current: A strong downward-moving current along the continental margin caused by earthquakes or the settling of sediments.

Turbine: An energy-producing engine.

Tussocks: Small clumps of vegetation found in marshy tundra areas.

Typhoon: A violent tropical storm that begins over the ocean.

U

Understory: A layer of shorter, shade-tolerant trees that grow under the forest canopy.

Upstream: The direction from which a river or stream is flowing.

Upwelling: The rising of water or molten rock from one level to another.

V

Veld: Temperate grassland in South Africa.

Venom: Poison produced by animals such as certain snakes and spiders.

Vertebrates: Animals with a backbone.

W

Wadi: The dry bed of a stream that flows only after rain; also called a wash or an *arroyo*.

Warm-bodied fish: Fish that can maintain a certain body temperature by means of a special circulatory system.

Wash: The dry bed of stream that flows only after rain; also called an *arroyo* or a *wadi*.

Water column: All of the waters of the ocean, exclusive of the sea bed or other landforms.

Water cycle: Natural cycle in which trees help prevent water runoff, absorb water through their roots, and release moisture into the atmosphere through their leaves.

Waterfall: A cascade of water created when a river or stream falls over a cliff or erodes its channel to such an extent that a steep drop occurs.

Water table: The level of groundwater.

Waves: Rhythmic rising and falling movements in the water.

Westerlies: Winds occurring between 30 degrees and 60 degrees latitude; they blow in a westerly direction.

Wet/dry cycle: A period during which wetland soil is wet or flooded followed by a period during which the soil is dry.

Wetlands: Areas that are covered or soaked by ground or surface water often enough and long enough to support plants adapted for life under those conditions.

Wet meadows: Freshwater marshes that frequently dry up.

X

Xerophytes: Plants adapted to life in dry habitats or in areas like salt marshes or bogs.

Y

Young stream: A stream close to its headwaters that has a narrow channel with steep banks.

Z

Zooplankton: Animals, such as jellyfish, corals, and sea anemones, that float freely in ocean water.

RIVER AND STREAM

A river is a natural flow of running water that follows a well-defined, permanent path, usually within a valley. A stream (also called a brook or a creek) is a natural flow of water that follows a more temporary path that is usually not in a valley. However, the term *stream* is often used to mean any natural flow of water, including rivers. Although some rivers are larger than some streams, size is not a distinguishing factor.

The origin of a river or stream is called its source. If its source consists of many smaller streams coming from the same region, they are called its headwaters. Its channel is the path along which it flows, and its banks are its boundaries, the sloping land along each edge between which the water flows. The point at which a stream or river empties into a lake, a larger river, or an ocean, is its mouth. When one stream or river flows into another, usually larger, stream or river, and adds its flow, it is considered a tributary of that larger river. Many tributaries make up a river system. The vast Amazon system in South America, for example, is fed by at least 1,000 tributaries.

The world's longest river is the Nile in eastern Africa. It begins in Ethiopia and travels for 4,157 miles (6,651 kilometers) into Egypt, where it discharges its waters into the Mediterranean Sea. The river system that obtains water from the largest area—2,722,000 square miles (7,077,200 kilometers)—is that of the Amazon. The Amazon is also the river that transports the largest volume—20 percent of all the water in all the rivers of the world.

HOW RIVERS AND STREAMS DEVELOP

Rivers and streams are part of the Earth's hydrologic cycle. The hydrologic cycle describes the manner in which molecules of water evaporate, condense and form clouds, and return to the Earth as precipitation (rain, sleet, or snow). Rivers pass through several stages of development.

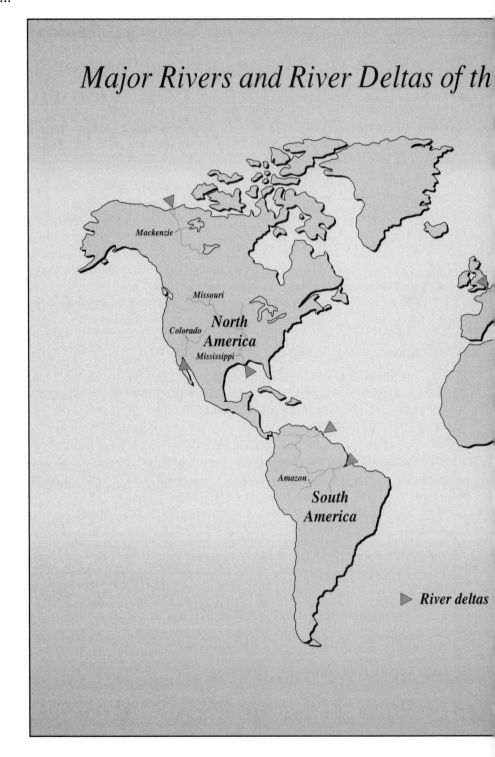

Major Rivers and River Deltas of th

Mackenzie

Missouri

North America

Colorado

Mississippi

Amazon

South America

▶ River deltas

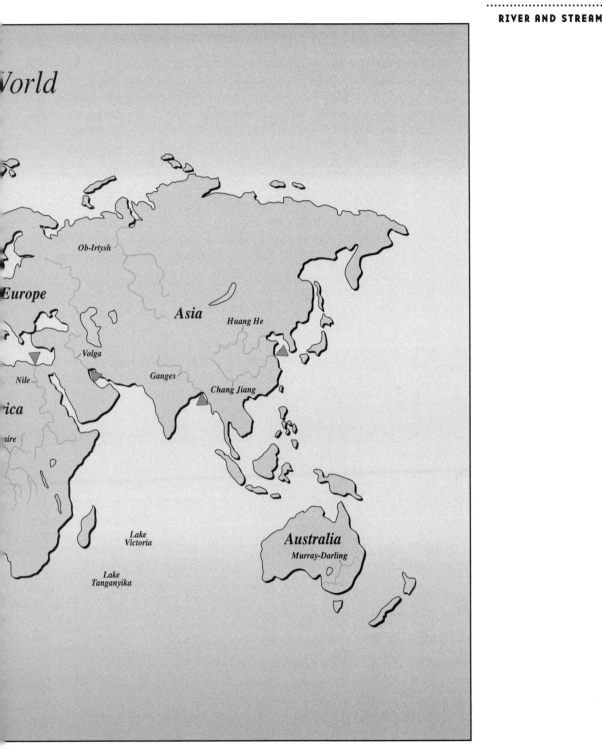

World

Ob-Irtysh

Europe

Asia

Huang He

Volga

Nile

Ganges

Chang Jiang

rica

aire

Lake
Victoria

Australia
Murray-Darling

Lake
Tanganyika

FORMATION

Rivers and streams owe their existence to precipitation, lakes, and groundwater, combined with gravity and a sloping terrain.

When rain falls on the land, often the soil cannot absorb it all. Much runs off and with the aid of gravity travels downhill, creating rills (tiny gullies). Many of these rills may meet at some point and their waters run together to form bigger gullies until eventually all this water reaches a valley or gouges out its own large channel. When enough water is available to maintain a steady ongoing flow, a stream or river is the result. Gravity and the pressure of the flowing water cause the river to travel until it is either blocked, in which case the water backs up and forms a lake, or empties into an existing lake or ocean.

Most of the precipitation that feeds streams and rivers comes from runoff. However, precipitation may also be stored as ice in glaciers in arctic regions or on mountaintops. As the glaciers melt, they nourish streams, and the streams feed rivers. The Rhine River in Germany, for example, obtains much of its water from glaciers in the Swiss Alps.

A lake can also be a source of river water. If the land slopes away from the lake at some point and the water level is high enough for it to overflow, a river or stream may form.

Another source of river water is groundwater. Groundwater is water that has seeped beneath the Earth's surface where it becomes trapped in layers of rock called aquifers. The Ogallala Aquifer under the Great Plains in the United States is an example. When an aquifer is full, its water escapes to the

MAJOR RIVER SYSTEMS

Name	Location	Length (mi./km.)	Basin Area (sq. mi./sq. km.)
Nile	North Africa	4,157/6,651	1,100,000/2,860,000
Amazon	South America	3,915/6,437	2,722,000/7,077,200
Mississippi-Missouri	North America	3,860/6,176	1,243,700/3,233,620
Ob-Irtysh	Siberia	3,461/5,538	959,500/1,919,000
Yangtze	China	3,434/5,494	756,498/1,966,895
Murray-Darling	Australia	3,371/5,394	414,253/1,077,058
Yenisei-Angara	Siberia	3,100/4,960	1,003,000/2,607,800
Zaire-Congo	Africa	2,716/4,346	1,425,000/3,705,000

Stages in River Development

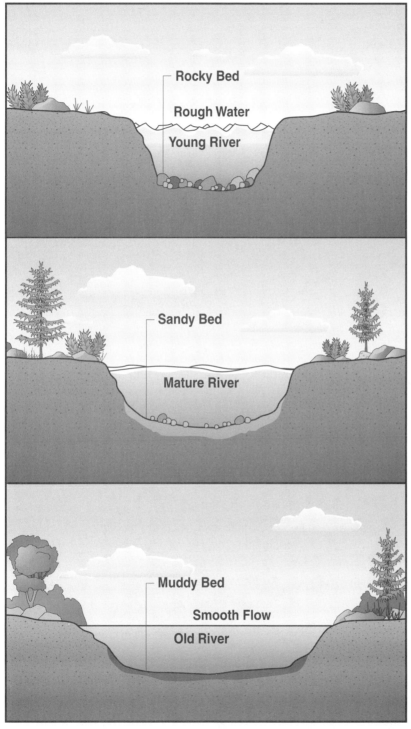

Rocky Bed

Rough Water

Young River

Sandy Bed

Mature River

Muddy Bed

Smooth Flow

Old River

An illustration of the stages in a river's development showing a young, mature, and old river.

surface, either by seeping directly into a river or stream bed or by forming a spring (an outpouring of water), which may then become a river's source. As much as 30 percent of the water in rivers and streams comes from groundwater. It is estimated, for example, that groundwater supplies about half of the water in the Mississippi River. However, the opposite may also happen. River water may seep into the ground and fill an aquifer, as the Colorado River does as it travels through Arizona, Nevada, and California.

STAGES OF DEVELOPMENT

Rivers are said to "age," not in terms of how long they have existed but in terms of their development as they travel across the land. They are either young, mature, or old. Some rivers, such as the Mississippi, may be in all three stages of development at one time.

Young rivers usually occur in highland or mountainous regions and have narrow, rocky channels with many boulders. The water may form waterfalls or foam and gurgle as it rushes over the rocks. Young rivers have few tributaries.

Mature rivers are those that have reached flat land as their tributaries pour more water into them. They often become flooded during periods of heavy rain or snowmelt. Their beds tend to be muddy rather than rocky because of the sediment (particles of sand or soil) carried into them by swift-flowing streams, and the valleys through which they flow are usually fairly broad. Few waterfalls occur on mature rivers, but they may have many bends or loops as they curl across the land looking for the lowest level to follow.

A river is old near its mouth, where layers of sediment build up over time. Here, the land is wide and flat, and the water travels more smoothly than in young and mature rivers.

KINDS OF RIVERS AND STREAMS

Rivers and streams can be classified according to their degree of permanence, the shape of their channels, and their branching network.

DEGREE OF PERMANENCE

Permanent streams and rivers flow all year long. Enough water is available to keep them from drying up completely, even during a long, dry spell. Most large rivers are permanent.

Intermittent streams and rivers are seasonal. They occur only during the rainy season or in the spring after the snow melts. Rivers and streams in desert regions tend to be intermittent, where they are also called *wadis* or *arroyos*.

Interrupted streams and rivers flow above ground in some places and then disappear from sight as they dip down under sand and gravel to flow underground in other places. The Santa Fe River in Florida is an example of an interrupted river.

CHANNEL SHAPE

The material over which a stream or river flows, and the force of the water as it travels determine the shape of the channel, which can be straight, braided, or meandering.

A straight stream or river flows in a straight line. This type is very rare, because flowing water tends to trace a weaving path. Those straight streams that do occur tend to have rocky channels.

A braided stream or river, such as the Platte River in Nebraska, consists of a network of interconnecting channels broken by islands or ridges of sediment, primarily mud, sand or gravel. Braided streams often occur in highland regions and have a steep slope.

A meandering stream or river winds snake-like through relatively flat countryside. Meandering streams tend to have low slopes and soft channels of silt (soil) or clay. The Menderes River in Turkey meanders, and the term, which means to wander aimlessly, is derived from its name.

ARCHITECTS OF THE UNDERGROUND

In areas where large deposits of limestone and other soft rock are found, underground rivers may create vast networks of caves. The best example is Mammoth Cave in Kentucky, which has large chambers and underground passageways on five levels that wind back and forth for a total length of 150 miles (240 kilometers). Several underground rivers and streams still flow through the cave, including Echo River, on which tour boats are operated. The beautiful caverns and rock formations carved by the rivers have been given intriguing names, such as King Solomon's Temple, the Pillars of Hercules, and the Giant's Chamber. Ancient Indian tribes once used the caves, and their mummified bodies have been found there.

BRANCHING NETWORK

Rivers do not usually originate from a single spot but are the result of a branching network that resembles the branches on a tree. The smallest streams or branches that do not have tributaries are called first-generation streams. If two first-generation streams join to form a new stream, the new stream is considered a second-generation stream. When two second-generation streams join, they form a third-generation stream, and so on. A very large river the size of the Mississippi is usually considered a tenth-generation stream.

THE WATER COLUMN

The water column refers to just the water in a river or stream, exclusive of its channel (path) or banks. All the world's rivers and streams combined contain less than 1 percent of all the water on Earth. Most water is held in the oceans.

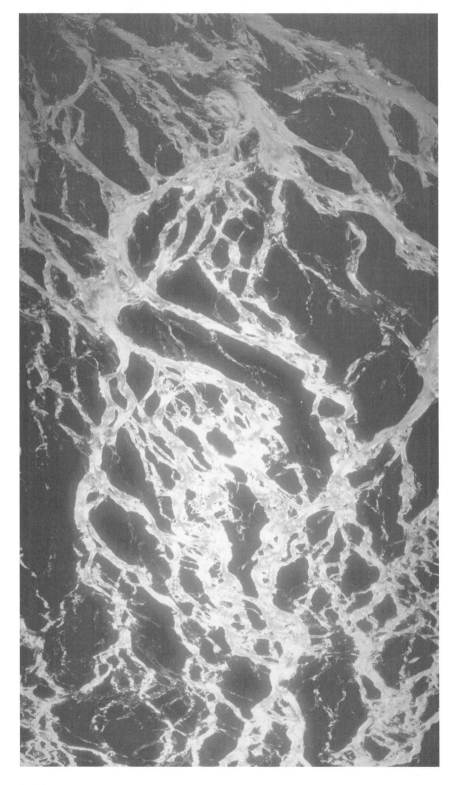

Braided streams of the
Firth River in Canada
reflecting the sunlight.
(Reproduced by
permission of Corbis.
Photograph by Lowell
Georgia.)

COMPOSITION

Rivers and streams carry fresh water, but this water may be clear or cloudy, polluted or clean. What determines its characteristics are where it originates and the nature of its channel.

As the water moves across the land, particles of rock, soil, and decaying plant or animal matter, called sediment, become suspended in it. An average of 0.13 ounces per gallon (100 milligrams per liter) of material is suspended in rivers in humid regions of the United States, and as much as 133 ounces per gallon (100,000 milligrams per liter) in rivers in desert regions. If a river carries a large quantity of sediment, it will be muddy. The smaller the amount of sediment, the clearer the water.

River water also reacts chemically with the rocks in its channel and dissolves some of the minerals they contain, which are called "salts." These dissolved salts give the water its taste and make it "hard." Rivers and streams carry large amounts of these salts. The Niagara River, for example, which lies on the border between the state of New York and Ontario, Canada, carries an average of 60 tons (54 metric tons) of dissolved mineral salts per minute as it pours over Niagara Falls.

> ### MARK TWAIN
> Samuel Clemens (1835–1910), one of America's greatest authors, wrote under the name Mark Twain. "Twain" means "two fathoms deep" (a fathom is about 6 feet), and it was one of the terms shouted by crewmen on Mississippi river boats who were measuring the river's depth to help the captain know if the boat could pass through that part of the channel. Clemens himself worked for a time as a steamboat pilot, and many of his stories have to do with the river. His river adventures are chronicled in *Life on the Mississippi,* and his best work is considered to be *The Adventures of Huckleberry Finn,* in which the river also plays an important part.

CIRCULATION

Water in a river or stream is almost always in motion. The direction from which it comes is called upstream, and the direction toward which it flows is called downstream. If you paddle upstream, you ascend the river. If you paddle downstream, you descend.

Velocity Rivers and streams acquire their initial energy from their elevation. Their waters are always falling downhill toward their ultimate goal, sea level (the average level of the surface of the sea). The fastest velocities (speeds) occur at waterfalls. The water flowing over Niagara Falls, for example, has been clocked at 68 miles (109 kilometers) per hour. However, most rivers and streams seldom exceed 10 miles (1.6 kilometers) per hour, and the average speed is 2 to 4 miles (3.2 to 6.4 kilometers) per hour.

Although gravity helps get things started, gushing mountain streams actually move more slowly than large, broad, mature rivers. This happens because friction (resistance to motion when one object rubs against another) caused by water molecules rubbing against the channel and banks decreases downstream where the channel is usually smoother and less rocky. Also, where the river is wider and deeper and the volume of water is greater, a smaller percentage of water molecules are exposed to friction.

The water at different locations or levels in the same section of river may move at varying speeds. The velocity is fastest in the middle and just below the surface. Movement slows down with depth and along the banks because friction increases. Along the very bottom of the channel is the boundary layer, a layer of water only 0.1 inch (0.25 centimeter) deep, where friction has stopped the flow completely.

Wave and current Waves are rhythmic rising and falling movements in the water. Surface waves are caused by wind blowing across a river's surface. Winds tend to follow a regular pattern. They occur in the same place and blow in the same direction, and the movement of waves follows this pattern.

The current is the steady flow of the water in a particular direction, usually from upstream to downstream. But river currents can be influenced by the slope and composition of the channel and the position of landmasses. When a current travels down a very steep slope, it may gain speed and force. When it meets a landmass, such as an island, it may be deflected and turn in a new direction.

An eddy is a current that moves against the regular current, usually in a circular, or whirlpool, motion. A riffle is a stretch of rapid, shallow, or choppy water that alternates with a quiet pool. A riffle is usually caused as the current flows over stones or gravel. A quiet pool is a deep, still area of water.

Discharge A river's discharge is the amount of water that flows out of it over a given period of time into another river, a lake, or the ocean. All of the world's rivers and streams combined discharge a total of 28,000,000,000 gallons (105,980,000,000 liters) of water every day. If all this water were to flow out over the land, it would cover the land to a depth of 9 inches (25 centimeters).

Discharge is usually measured by multiplying the river's width at the surface by its average depth times its velocity. The Amazon, which empties into the Atlantic Ocean, discharges more water than any other river—more than 4,000,000 cubic feet (112,000 cubic meters) per second.

Tidal bores Tidal bores are surges of ocean water caused when ridges of sand direct the ocean's flow back into a narrow river channel, sometimes as a single wave. Most tidal bores are harmless, but in the Tsientang River in China, the bore sometimes reaches 25 feet (7.6 meters) in height.

THERE'S GOLD IN THEM THAR RIVERS!

Between 1860 and 1910, people stampeded to the Yukon Territory of Canada in search of gold. They found it not only in the ground, but also in the sediments of rivers, especially the Klondike, the Lewes, the Stewart, and the lower Yukon. These sediments were scooped up into a flat pan and sloshed with water in an effort to separate bits of rock and other debris from the shining gold particles. It is estimated that about $100,000,000 in gold was taken from the Yukon Territory during that time.

FLOODS

Floods are caused when more water enters the river than its channel can hold. The result is increased discharge and high water levels. Most rivers overflow their banks every two to three years, although the amount of overflow is usually moderate. Depending on the river's origins and location, flooding is caused by melting snow, melting glaciers, or heavy rainfall, and sometimes by all three. Rivers supplied by snow or glaciers ordinarily produce only moderate flooding, because snow and glaciers melt slowly and the extra water does not enter the river all at once. When heavy rainfall is added, however, the discharge may be enormous. Because rivers can be long, rain in a location upstream may cause a flood downstream in a region that received no rain at all.

The seriousness of a flood is measured by comparing the river's average annual discharge rate to the rate during the flood. The Huang He (Yellow) River in China, for example, has an average discharge rate of 111,818 cubic feet (3,131 cubic meters) per second. During a flood in 1843, its rate increased to 1,230,000 cubic feet (34,440 cubic meters), which was 11 times greater than normal. Heavy floods can also cover large areas. In 1954, the Chang Jiang (Yangtze) River in China flooded several towns and lowlands near Hankow for an area 50 miles (80 kilometers) wide and 100 miles (160 kilometers) long.

> ## HOW THE AMAZON RIVER GOT ITS NAME
>
> The first European to explore the Amazon and its basin was Francisco de Orellana of Spain (c. 1511–1546). In 1541, he made a 17-month trip from the foot of the Andes Mountains in the west to the mouth of the Amazon River in the east. The journey was a difficult one with many battles with the Indians. At times food was scarce and he and his men had to eat toads and snakes. This lack of food may partly explain why Orellana thought they had also been attacked by a band of female warriors resembling the famous Amazons of Greek legends. Although other explorers later failed to encounter these hostile females, Orellana had named the river in their honor, and the name stuck.

In very dry regions, a river or stream may dry up completely for several weeks or months. During a sudden rainstorm, the channel may not be large enough to contain the amount of water, and a flash flood may occur. Flash floods are dangerous because they happen very suddenly, the wall of water bringing everything down before it.

EFFECT ON CLIMATE AND ATMOSPHERE

The climate of a stream or river depends upon its location as it travels over land. In general, if it passes through a desert country, such as Israel, the climate will be hot and dry. If it begins on top of a mountain in the Northern Hemisphere, the climate will be cold. Rivers and streams in temperate (moderate) climates are often affected by seasonal changes. In these areas, small streams or rivers may dry up in summer or develop a layer of ice in winter.

The presence of a large river can itself create some climatic differences. In summer, evaporation may create more moisture in the air. When the water temperature is cooler than the air temperature, winds off the river can help cool the nearby region. The water in rivers and streams is also partly responsible for the precipitation that falls on land. The water evaporates in the heat of the Sun, forms clouds, and falls elsewhere.

GEOGRAPHY OF RIVERS AND STREAMS

Rivers and streams are strong forces in shaping the landscape through which they flow. As the current moves against the channel and banks, the particles of sediment the river carries wear away the surface with a cutting action called erosion (ee-ROH-zuhn). All the world's rivers combined remove an average of 2 tons (1.8 metric tons) of rock and soil from every square mile (2.6 square kilometers) of land they cross each year. Some chunks can be huge. During a flood in 1923, a stream in the Wasatch Range of mountains in Idaho and Utah picked up boulders weighing as much as 90 tons (82 metric tons) and carried them more than 1 mile (1.6 kilometers) downstream.

The faster a river flows, the faster it wears the land away and the more sediment it bears. Some of the eroded chunks and particles may sink to the bottom. Others are carried along and, as the river slows down, are dropped farther downstream. Erosion (wearing away) and deposition (dep-oh-ZIH-shun; layering) make many changes over time.

THE DRAINAGE BASIN

A river or stream obtains runoff from an area of land called its drainage basin, or watershed. The drainage basin can be extremely large. The Mississippi-Missouri river system, for example, has as its drainage basin the entire area between the Appalachian Mountains in the east and the Rocky Mountains in the west. The world's largest drainage basin is that of the Amazon, which measures 4,200,000 square miles (10,920,000 square kilometers).

CHANNEL AND BANKS

The channel of a river or stream slopes down to the center, where it is deepest. The bottom of the channel is called its bed. Channel beds may be rocky or covered with gravel, pebbles, or mud. Some river bed sediments are formed from the waste products and dead tissues of animals and plants. Others usually consist of clay, stone, and other minerals.

The general shape of the banks is partly determined by the general shape of the surrounding land. If the river runs through a plain, then its banks will be broad and level. If it occurs in the mountains, then its banks may be steep and rocky.

As water rushes through the channel of a young river, erosion tends to scour it out and make it deeper. Young river valleys are often narrow and steep sided. As the river matures and velocity and volume increase, width tends to increase more than depth. A mature valley tends to have a gentle slope, and an old valley may be almost flat.

Moderate periodic flooding is a major cause of changes in bank shape. Although large floods cause more erosion, they do not occur often enough to make drastic or permanent changes.

LANDFORMS

Landforms created by rivers and streams include floodplains and levees; wetlands; canyons and gorges; rapids and waterfalls; spits, bars, and shoals; and deltas and estuaries.

Gooseneck Canyon in Utah was cut into the limestone by the San Juan River over millions of years. (Reproduced by permission of Corbis. Photograph by Tom Bean.)

Floodplains and levees When a stream carrying a large quantity of sediment flows into another stream or out over a plain, it slows down and its sediments settle to the bottom. These sediments (called alluvium) often spread out in a fan shape, creating what is called an alluvial fan. When this

occurs frequently, the result may be a floodplain. A floodplain is a flat region with soil enriched by river sediments and usually makes good farmland.

Very silty rivers may create high banks of sediments called levees. The levees of the Mississippi River are several yards (meters) high in places.

Wetlands Sometimes, during a flood, a river may change its path. The current may seek out a new path or, as the flood waters recede, sediments prevent the water in a loop or bend from rejoining the rest of the river. Here, a wetland, such as a marsh or swamp, or an "oxbow" lake may form. Oxbow lakes have a curved shape, which gives a clue to the origin of their name (An oxbow is an u-shaped piece of wood placed under an ox.)

Canyons and gorges Many riverbanks consist of both hard and soft rock. Moving water erodes the soft rock first, sometimes sculpturing strange shapes. In regions where much of the rock is soft, rivers and streams may cut deep canyons (long narrow valleys between high cliffs) and gorges (deep, narrow passes). The Gooseneck Canyon in Utah was cut into the soft limestone of the region by the San Juan River over millions of years. The distance from the top of the canyon to where the river is now is 1,500 feet (456 meters).

Caves may be gradually carved into the sides of cliffs by erosion. In large rivers, headlands may be created. A headland is an arm of land made of hard rock that juts out from the bank into the water after softer rock has been eroded away.

Rapids and waterfalls When fast-moving water erodes softer rock downstream, gradually cutting away the bed in places and creating many short drops, rapids may form. Rapids are often called "white water" because of the foam created when the rushing water hits the exposed bands of rock. The Salmon River in Idaho flows through steep canyons and has many rapids. Its rapids are so dangerous to travelers that it has been nicknamed the "River of No Return."

When a river or stream falls over a cliff or erodes the channel to such an extent that a steep drop occurs, it creates a waterfall. The world's highest waterfall is Angel Falls on the Rio Churun in southeastern Venezuela. Its greatest uninterrupted drop is 2,640 feet (800 meters). Victoria Falls, on the Zambezi River between Zimbabwe and Zambia, is considered one of the Seven Natural Wonders of the World. It is the world's largest falls—more than 4,400 feet (1,341 meters) wide—and Africans call it *Mosi-oa-tunya*, the "smoke that thunders."

Where a waterfall strikes the valley below, it gouges out a deep basin called the plunge pool. Eventually, it also erodes the lip of rock over which it flows. As erosion continues over time, the waterfall slowly moves upstream. Ten thousand years ago, Niagara Falls, for example, was almost 7 miles (11 kilometers) farther downstream than it is now.

Spits, bars, and shoals A spit is a long narrow point of deposited sand, mud, or gravel that extends into the water. A bar is an underwater ridge of

OPPOSITE:
Angel Falls, the highest waterfall in the world, was discovered in 1935 and named after American pilot James Angel who crashed his plane in the area. (Reproduced by permission of Corbis. Photograph by Pablo Corral V.)

sand or gravel, formed by currents, that extends across a channel or the inside bank of a curve. Shoals are areas where enough sediments have accumulated that the water is very shallow and dangerous for navigation.

Deltas and estuaries Where rivers meet a lake or ocean, huge amounts of silt can be deposited along the shoreline. Large rivers can dump so much silt that islands of mud build up, eventually forming a triangle-shaped area called a delta. The finer debris that does not settle as quickly may drift around, making the water cloudy. The Nile River in Egypt and the Ganges (GANN-jeez) in India both have large deltas. That of the Ganges covers 23,166 square miles (60,000 square kilometers).

When a river traveling through lowlands meets the ocean in a semi-enclosed channel or bay, the area is called an estuary. The water in an estuary is brackish—a mixture of fresh and salt water. In these gently sloping areas, river sediments collect, and muddy shores form. Chesapeake Bay along the coast of Maryland is the largest estuary on the Atlantic coast of the United States. Estuaries are often sites for harbors because they can serve both river and ocean travel.

Men stand talking and bathing in the Ganges River in India. (Reproduced by permission of Corbis. Photograph by Ric Ergenbright.)

ELEVATION

Rivers and streams occur at all altitudes, and their altitude changes over their length because they usually begin in highlands or mountains and move to lower elevations. The headwaters of the Ganges River in India, for example, issue from an ice cave in the Himalaya Mountains at an elevation of 10,300 feet (16,480 meters). A thousand miles (1,600 kilometers) from its mouth, the Ganges is only about 700 feet (213 meters) above sea level, and 0 feet (0 meters) at its mouth where it enters the Bay of Bengal.

PLANT LIFE

Few plants can root and grow in running water; therefore, most of those that live in rivers and streams are found along the banks or in quiet pools where the environment is similar to that of a lake or pond. Plants with floating leaves do not do well in fast-moving streams because the current pulls the leaves under. But when rooted plants gain a foothold and manage to reproduce, they can grow so numerous that they slow the flow of water and can even cause flooding.

River and stream plants may be classified as submergent, floating aquatic (water), or emergent according to their relationship with the water.

A submergent plant grows beneath the water. Even its leaves lie below the surface. Submergents include the New Zealand pygmyweed, tape grass, and water violet.

Floating aquatics float on the water's surface. Some, such as duckweed, have no roots. These are also called phytoplankton (fi-toh-PLANK-tuhn; "plankton" in Greek means "wanderer"). Others, such as the water lily, have leaves that float on the surface, stems that are underwater, and roots that are anchored to the bottom.

An emergent plant grows partly in and partly out of the water. The roots are usually underwater, but the stems and leaves are at least partially exposed to air. They have narrow, broad leaves, and some even produce flowers. Emergents include reeds, rushes, grasses, cattails, and sedges.

Plants that live in rivers and streams can be divided into four main groups: bacteria; algae (AL-jee); fungi (FUHN-jee) and lichens (LY-kens); and green plants.

A young man gathers rushes from the edge of a river in East Anglia, England. (Reproduced by permission of Corbis. Photograph by Adam Woolfitt.)

ALGAE

Most submergent plants are algae. (Although it is generally recognized that algae do not fit neatly into the plant category, in this chapter they will be discussed as if they were plants.) Some forms of algae are so tiny they cannot be seen without the help of a microscope. These, like spirogyra, sometimes float on the water in slow-moving rivers or coat the surface of rocks, other plants, and river debris. Other species, such as blanket weed, are larger and remain anchored to the riverbed in quiet pools.

Certain types of algae have the ability to make their own food by photosynthesis (foh-toh-SIN-thih-sihs), the process by which plants use energy from sunlight to change water and carbon dioxide from the air into the sugars and starches they use for food. A by-product of photosynthesis is oxygen, which combines with the water and enables aquatic animals, such as fish, to "breathe." Algae also require other nutrients that must be found in the water, such as nitrogen, phosphorus, and silicon. Algal growth increases when nitrogen and phosphorus are added to a stream by sewage or by runoff from fertilized farmland.

Growing season Algae contain chlorophyll, a green pigment used during photosynthesis. As long as light is available, algae can grow. But growth is often seasonal. In some areas, such as the Northern Hemisphere, the most growth occurs during the summer months when the Sun is more nearly overhead. In temperate zones, growth peaks in the spring but continues throughout the summer. In regions near the equator, no growth peaks occur, since it is warm year round. As a result, growth is steady throughout the year.

Reproduction Algae may reproduce in one of three ways. Some split into two or more parts, each part becoming a new, separate plant. Others form spores (single cells that have the ability to grow into a new organism), and a few reproduce sexually, during which cells from two different plants unite to create a new plant.

Common river and stream algae Two types of algae are commonly found in rivers and streams: phytoplankters and macrophytic algae.

PHYTOPLANKTERS Phytoplankters float on the surface of the water. Two forms of phytoplankton, diatoms and dinoflagellates (dee-noh-FLAJ-uh-lates), are the most common. Diatoms have simple, geometric shapes and hard, glasslike cell walls. They live in colder regions and even within arctic ice. Dinoflagellates have two whiplike attachments that make a swirling motion. They live in tropical regions (regions around the equator).

MACROPHYTIC ALGAE Macrophytic algae—"macro" means "large"—usually grow attached to the bottom of the channel. Some have balloon-like, air-filled swellings at the bases of their leaflike fronds (branches), which help them remain upright. Although they resemble green plants, they have no true leaves, stems, or roots. Instead they have a rootlike structure, called a holdfast, that anchors them to the bed.

Macrophytic algae are rare in fast-moving water, although some species can be found growing on stable surfaces.

FUNGI AND LICHENS

As with algae, it is generally recognized that fungi and lichens do not fit neatly into the plant category. Fungi are plantlike organisms that cannot make their own food by means of photosynthesis. Instead, they grow on decaying organic (material derived from living organisms) matter or live as parasites (organisms that depend upon other organisms for food or other needs) on a host. Fungi grow best in a damp environment and are often found along river-banks. Common fungi include mushrooms, rusts, and puffballs.

Lichens are combinations of algae and fungi that tend to grow on rocks and other smooth surfaces. The alga produces the food for both itself and the fungus by means of photosynthesis. The fungi may provide moisture for the algae. One of the most common lichens found near streams is called reindeer moss, which prefers the solid surfaces of rocks.

GREEN PLANTS

The true green plants found growing in the water along the banks of rivers and streams are similar to those that grow on dry land. Unlike algae, they have roots and some, like the hornwort, may even bloom underwater. Large beds of green plants slow the movement of water and help prevent erosion. Some water animals use green plants for food and for hiding places. Those plants, such as sallows (European species of willows) and sedges, that occupy the emergent zone between the water-covered area and dry land need moist, but not saturated, soil.

Most green plants need several basic things to grow: light, air, water, warmth, and nutrients. Near a stream, light and water are in plentiful supply. Nutrients—primarily nitrogen, phosphorus, and potassium—are usually obtained from the soil. Some soils, however, are lower in these nutrients. They may also be low in oxygen. These deficiencies may help limit the kinds of plants that can grow in an area.

Growing Season Climate and the amount of precipitation both affect the length of the growing season. Warmer temperatures and moisture usually signify the beginning of growth. In regions that are colder or receive little rainfall, the growing season is short. Growing conditions are also affected by the amount of moisture in the soil, which can range from saturated during flooding or a rainy season, to dry.

Reproduction Green plants reproduce by several methods. One is pollination, in which the pollen from the male reproductive part (called a stamen) of a plant is carried by wind or insects to the female reproductive part (called a pistil). Water lilies, for example, which are closed in the morning and evening, open during midday when the weather is warmer and insects are more active and more likely to visit. Some shoreline plants, like sweet flag, reproduce by sending out rhizomes (RY-zohms), which are stems that spread out under water or soil and form new plants. Reed mace and common reeds reproduce by this means.

Common river and stream green plants Common plants found in and around rivers and streams include willow moss, water lily, pondweed, duckweed, aquatic buttercup, great pond sedge, water plantain, water lettuce, azoilla water fern, water soldier, bur reed, marsh horsetail, reed mace, mare's tail, greater spearwort, flowering rush, water forget-me-not, St. John's wort, alder, and weeping willow.

ENDANGERED SPECIES

Changes in the habitat, such as pollution and the presence of dams, endanger river and stream plants. The plants are also endangered by people who collect them, and by the use of fertilizers and herbicides (poisons used to control weeds), which enter the water as runoff from farms.

The papyrus plant in Egypt, for example, is endangered because of the Aswan High Dam. The dam traps Nile water in its reservoir upstream, and the swamps and ponds that form the habitat for papyrus are disappearing.

ANIMAL LIFE

In addition to the land animals, such as raccoons that visit for food and water, rivers and streams support many species of aquatic animals. Some swim freely in the water, while others live along the muddy bottom. Some prefer life in midstream or on rocks beneath a fast-moving flow; others seek the shallows or quiet pools. Shallows are often warm and exposed to sunlight, but animals that prefer cool shady spots can find them along the banks beneath overhanging trees. Many river and stream animals are adapted to life under high-speed conditions. Salmon and trout, for example, are torpedo-shaped and can swim against the current more easily than some other fish. Some, like torrent beetles and stonefly larvae, have low-slung or flattened bodies that enable them to cling to rocks without being washed away.

Water quality often determines which species will be supported in a particular river or stream. Temperature, velocity, oxygen content, mineral content, and muddiness are all factors. Cold-water trout, for example, prefer cool, shady streams, while snails require calcium-rich waters in order to build their shells. Because they are a combination of both salt- and freshwater environments, estuaries are home to species especially adapted to those conditions. (For more information about estuarine animals, see the chapters titled "Continental Margin" and "Seashore.")

MICROORGANISMS

Microorganisms cannot be seen by the human eye without the use of a microscope. Those found in rivers and streams include the transparent stentor. Also, the larvae of animals whose adult forms may be larger, such as dragonflies, worms, and frogs, may spend some time as microorganisms that live in the water.

Bacteria Bacteria are microorganisms found throughout rivers and streams and make up much of the dissolved matter in the water column. As such, they provide food for lower animals. They also help decompose (break down) the dead bodies of larger organisms, and, for that reason, their numbers increase along the banks where the most life is found.

INVERTEBRATES

Invertebrates are animals without a backbone. They range from simple flatworms to more complex animals such as spiders and snails.

Many species of insects live in rivers and streams. Some, like the diving beetle, spend their entire lives in the water. Others, like black flies, live in the water while young but leave it when they become adults.

Mollusks, such as freshwater mussels, are invertebrates with a hard outer shell that often inhabit rivers.

Food Insects may feed underwater as well as on the surface. They may be herbivores (plant eaters), carnivores (meat eaters), or scavengers that eat decaying matter.

The diets of invertebrates other than insects also vary. Some snails are plant feeders that eat algae, while freshwater crabs are often omnivorous, eating both plants and animals.

Reproduction Most invertebrates have a four-part life cycle. The first stage is spent as an egg. The second stage is the larva, which may actually be divided into several stages between which there is a shedding of the outer skin casing. A caterpillar is an example of an invertebrate in the larval stage. The third stage is the pupal stage, during which the insect lives in yet another protective casing, like a cocoon. Finally, the adult breaks through the casing and emerges.

For some insects, their entire life cycle is centered on breeding. Mayflies, for example, live for an entire year in streams as larvae. After they shed the larval casing, they breed, lay eggs, and die—all within a day or two.

If enough mayflies survive the larval stage, the numbers of dead adults can be enormous. On one occasion, a snowplow had to be used to remove a layer of mayflies several feet thick from a bridge across the Mississippi.

RIVER BLINDNESS

In tropical Africa, South America, and Central America, a round worm causes a disease in humans called river blindness. The worm is introduced by means of the black fly, which thrives around fast-moving rivers. The fly carries the larvae of the worm and spreads them to humans in its bite. The larvae burrow under the skin and spread throughout the body, eventually reaching the eyes. In some African villages, 15 percent of the people are infected.

Common river and stream invertebrates Common invertebrates found in rivers and streams include diving beetles, mayflies, water fleas, torrent beetles, stoneflies, dobsonflies, water boatmen, water striders, whirligig beetles, black flies, fisher spiders, water scorpions, water sticks, leeches, flatworms, snails, freshwater clams, freshwater mussels, raft spiders, caddisflies, and freshwater crayfish.

RAFT SPIDER Raft spiders live along the edges of slow-moving rivers, where they are able to run across the water's surface in search of insect prey. Bristles on their feet help distribute their weight so they don't sink. Raft spiders can be identified by two yellow stripes running the length of their bodies.

CADDISFLY Adult caddisflies are rather drab creatures that come out only at night, rarely feed, and live for only a couple of days. In their youth, however, they attract a lot of attention. The larvae of caddisflies use stones, sand, shells, and other items to build a protective shell around themselves. The only opening is a hole at the front through which the larva's head emerges to feed. Some larvae eat algae, and others build nets with which to catch food carried by the current.

FRESHWATER CRAYFISH The freshwater crayfish is found in rivers on all continental landmasses except Africa. It is related to the lobster and needs mineral-rich waters to build its protective shell. Species range from 1 inch (2.5 centimeters) to 16 inches (41 centimeters) in length. They seek shelter under stones in the stream bed or may even burrow into it, sometimes as far as 20 feet (6 meters) down. They catch prey, such as small fish, with their large pincer claws. Some blind, albino (colorless) species live in underground rivers.

AMPHIBIANS

Amphibians, including frogs, toads, newts, and salamanders, are vertebrates, which means they have a backbone. Amphibians live at least part of their lives in water. Because they breathe through their skin, and only moist skin can absorb oxygen, they must usually remain close to a water source. If they are dry for too long, they will die.

Amphibians are cold-blooded animals, which means their body temperatures are about the same temperature as their environment. As temperatures grow cooler, they slow down and seek shelter in order to be comfortable. In cold or temperate regions, some amphibians hibernate (remain inactive), digging themselves into the mud. When the weather gets too hot, many go through another similar period of inactivity called estivation.

Food In their larval form, amphibians are usually herbivorous, and the adults are usually carnivorous, feeding on insects, slugs, and worms. Those that live part of their lives on land have long, sticky tongues with which they capture their food.

Reproduction Most amphibians lay jelly-like eggs in the water. Frogs can lay up to as many as 3,000 eggs, which float beneath the water's surface. After being fertilized by the sperm of male frogs, the eggs hatch into tadpoles (larvae), and require a water habitat as they swim and breathe through gills. When the tadpole turns into a frog, it develops lungs and can live on land.

Spotted newts hatch in the water and live there as larvae, even developing gills. Later they lose their gills and live for a time on dry land. Two or three years later, they return to the water where they live the remainder of their lives.

Some amphibians, like the common frog, prefer quiet water for breeding. Others, like the large hellbender salamander, seek rushing streams.

Common river and stream amphibians The most common amphibians found in rivers and streams all over the world are salamanders, frogs, toads, and newts.

The palmate newt lives on land for part of the year where it hunts at night for worms and other small animals. In spring it returns to the water where it breeds and lays eggs. Most newts lay eggs one at a time and attach them to water plants. Some species even wrap each egg in a leaf for protection.

REPTILES

Reptiles are cold-blooded vertebrates that depend upon the environment for warmth. They are more active when the weather and water temperature become warmer. Many species of reptiles, including snakes, lizards, turtles, alligators, and crocodiles, live in temperate and tropical rivers and streams. All are air-breathing animals, with lungs instead of gills, that have adapted to life in the water. The Southeast Asian fishing snake, for example, can close its nostrils while swimming submerged.

Because they are so sensitive to the environment, many reptiles go through a period of hibernation in cold weather. Turtles, for example, bury themselves in the mud. They barely breathe and their energy comes from stored body fat. At the other extreme, when the weather becomes very hot and dry, some reptiles go through estivation.

Food All snakes are carnivores, and water snakes eat frogs, small fish, and crayfish. Some species of turtles are omnivores.

Reproduction The eggs of lizards and turtles are either hard or rubbery shells and do not dry out easily. Most are buried in warm ground, which helps them hatch. However, the eggs of a few species of lizards and snakes are held inside the female's body, and the babies are born alive.

Crocodiles keep their eggs warm in nests that can be simple holes in the ground or constructions above the ground made from leaves and branches.

Common river and stream reptiles Common reptiles found in rivers and streams include the water moccasin, the viperine water snake, the eastern water dragon, the soft-shelled turtle, the matamata turtle, the yellow-bellied terrapin, the snapping turtle, the anaconda, and the crocodile.

ANACONDA The anaconda of South American rivers is the world's largest snake, some individuals attaining lengths of more than 30 feet (10 meters). Although they are excellent swimmers and prefer to hunt along the edges of rivers and swamps, they can also climb trees. Anacondas are constrictors, killing their prey by coiling their body around it and suffocating it. Their young are born alive in the water.

CROCODILE Crocodiles hunt at twilight in tropical rivers. During the day they bask in the Sun at the river's edge. Then, as the light dims, they seek their prey. Some, such as the New Guinea crocodile, eat just fish. Most other species eat a variety of foods. They swallow smaller prey whole, but larger animals are dragged underwater where they drown and are torn apart and eaten.

FISH

Like amphibians and reptiles, fish are cold-blooded vertebrates. There are two types of freshwater fish. The first type, the parasites, which include some species of lampreys, attach themselves with suckers to other animals and suck their blood for food. The second type, which eat plant foods or catch prey, are the most numerous. They use fins for swimming and breathe with gills.

Many species of fish, such as archer fish, swim in schools. A school is a group of fish that swim together in a coordinated manner to discourage predators.

Food Some fish, such as the grass carp, eat plants while others, such as bluegills, depend upon insects, worms, and shellfish. Larger fish, such as the sturgeon, often eat smaller fish, and a few species feed on carrion (dead animals). Most fish specialize in what they eat and where they find their food. Some feed on the surface, and others seek food in deep water. Some prefer rushing streams and others calm pools. The arawana of Southeast Asian rivers, for example, has a large, upward-pointing mouth that helps it feed on the surface.

Reproduction Most fish lay eggs, and many species abandon the eggs once they are laid. Others build nests and care for the new offspring until they hatch. The male stickleback, for example, lures several females to his nest where they lay their eggs. He then fertilizes the eggs and guards them until they hatch. Still other species, such as certain catfish, carry the eggs with them until they hatch, usually in a special body cavity or even in their mouths.

Common river and stream fish Typical fish found in rivers and streams include the roach, rudd, tench, bream, perch, gudgeon, carp, bitterling, African knifefish, elephant fish, hatchetfish, shovel-nosed sturgeon, white sturgeon, paddlefish, freshwater shark, arapaima, stickleback, piranha, catfish, salmon, and trout.

PIRANHA Piranhas (purr-AHN-ahs) are predatory fish of South America. Having several rows of sharp, triangular teeth, they usually feed on other fish, or fruits and seeds that fall into the water. They become particularly dangerous during the dry season when the water level drops and the fish

A yellow anaconda hunting for food along a river's edge. Anacondas kill their prey by suffocating it. (Reproduced by permission of Corbis. Photograph by Joe McDonald.)

gather together in schools. A school of piranhas will attack and quickly devour large animals, including humans.

CATFISH Many species of catfish inhabit rivers worldwide. Some tropical varieties can climb out of the water and walk overland from one river or stream to another. They range in size from the small species that live in mountain streams, to those that live in European rivers and grow to a length of 10 feet (3 meters) and a weight of 500 pounds (227 kilograms).

Catfish got their name from the long feelers around the mouth that resemble cats' whiskers. They also have spines along the fins that release poison and cause painful injuries. Catfish that have adapted to life in fast-moving streams have fins with suckers or large, fleshy lips that help them cling to rocks and even climb. Bullheads, a species of catfish, have flattened bodies, which enables them to squeeze under stones and resist the current.

SALMON Salmon live in both fresh and salt water. The young hatch in fast-moving rivers and streams, where they live for about the first three years

A piranha, from the Amazon River in Brazil, bears its teeth. (Reproduced by permission of Corbis. Photograph by Wolfgang Kaehler.)

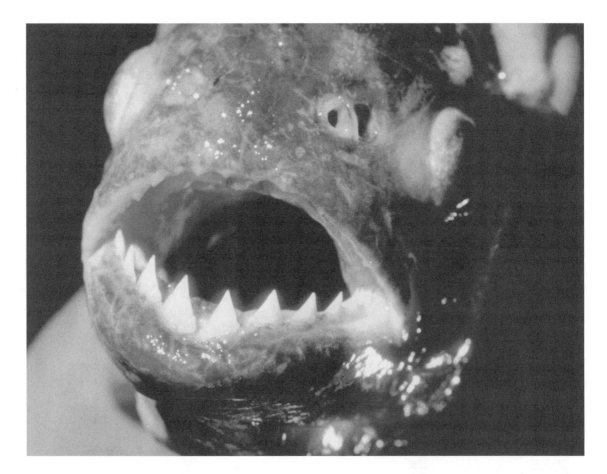

of their lives. They then migrate to the sea where they live for up to four years, feeding on small fish and shellfish. When it comes time to breed, they return to the river of their birth, swimming upstream over many obstacles such as rapids and waterfalls. After breeding, they die.

TROUT In general, trout are found in clear, cool, fast-moving streams with beds of gravel. They are carnivorous, feeding on insects, freshwater shrimp, and clams. Some species also prey on other fish. Rainbow trout, which originated in North America, have a silver belly and an iridescent stripe along the side.

BIRDS

Birds are vertebrates, and many different species live near rivers and streams. These include many varieties of wading birds, waterfowl, and shore birds. Most visit in search of food, and certain species use riverbanks as nesting places.

Although most aquatic birds prefer quiet waters, a few are specially adapted for the speed and turbulence of life upstream. One example is the torrent duck of the Andes Mountains in South America. Built like an ordinary duck in most respects, it has a longer tail, which helps it steer when swimming in fast currents, and claws on its webbed feet, which allow it to cling to rocks.

Food Nearly all birds must visit a source of fresh water each day to drink, and many feed on aquatic vegetation or the animals that live in the water. The dipper goes so far as to actually walk underwater in search of small animal prey.

One of the adaptations of birds to aquatic life is their beak, or bill. Some are shaped like daggers for stabbing prey such as frogs and fish, while others are designed to root through mud in search of food. Water-dwelling birds spend much time preening their feathers, for they depend on their feathers to keep them dry.

Reproduction All birds reproduce by laying eggs. Male birds are brightly colored, which attracts the attention of females. After mating, female birds lay their eggs in a variety of places and in nests made out of many different materials. One parent usually sits on the nest to keep the eggs warm while the other searches for food.

> ### ABSOLUTELY SHOCKING!
>
> There are probably lots of good reasons not to go swimming in tropical rivers, and one of them just has to be to avoid an electric shock. The African catfish and the Amazonian eel are two tropical river residents that produce their own electricity and use it as a weapon.
>
> The body of an electric eel is set up like a chemical battery. Thousands of electricity-producing cells are stacked in columns and oriented in the same direction. When the eel sights something tasty, a nerve impulse from the eel's brain triggers the flow of electric current from its tail, through the water, toward its head. When the eel touches its prey, the circuit is interrupted and the current flows into the body of the victim, stunning it. An adult eel can produce 600 volts of electricity, which is enough to knock out an animal the size of a horse—or a human.

Common river and stream birds Birds found around rivers and streams include kingfishers, green herons, blue herons, mergansers, reed warblers, reed buntings, African darters, African fish eagles, ospreys, sunbitterns, finfoots, egrets, and wood ducks. They fall into three basic categories: wading birds, shorebirds, and waterfowl.

WADING BIRDS Wading birds, such as herons, have long legs and wide feet for wading through shallow water. They also have long necks and bills necessary for nabbing food.

SHOREBIRDS Shorebirds, such as the sandpiper, feed and nest along the banks and prefer shallow water. Some, like the godwit, have long, slender, upturned bills, which help them sift through mud in search of food. Other birds, like the ruddy turnstone, have bills that are curved to one side which helps to overturn pebbles. The red knot is an example of shorebirds that have specialized cells in their bills that are "pressure sensitive," allowing them to detect shellfish buried beneath the sand.

WATERFOWL A waterfowl is a bird that spends most of its time on water, especially a swimming bird such as a duck. Their flattened bills are designed for grabbing the vegetation, such as sedges and grasses, on which they feed.

MAMMALS

Mammals are warm-blooded vertebrates covered with at least some hair, that bear live young. Aquatic mammals, such as the muskrat, often have fur that is waterproof. They may also have webbed toes for better swimming. The toes of the duck-billed platypus of Australia, for example, are webbed.

Food Certain aquatic mammals are carnivores. The water shrew, for example, eats insects, worms, and frogs. Others like muskrats are omnivores, while beavers are herbivores.

Reproduction Mammals give birth to live young that have developed inside the mother's body. Some mammals, such as otters, are helpless at birth, while others, like dolphins, are able to walk

WALKING ON WATER

The Jesus Christ lizard, or basilisk, of Central America got its name from its ability to walk on water. Scales and a flap of skin on its hind toes increase the surface area of its feet and enable it to scamper on the water's surface. Being lightweight doesn't hurt either!

Another water-walker, or at least that's what it appears to be, is the jacarana, or lily trotter, bird of the tropics. Although jacaranas have extremely long toes that distribute their weight over the surface, they are actually walking on the floating leaves of water plants.

THE FISH THAT CAN CLIMB LADDERS

In order to breed, Atlantic salmon return from the sea to rivers along the Atlantic and Pacific coasts where they first hatched. Unfortunately, most of these rivers have been dammed in places, preventing the fish from continuing on their journey upstream. To aid them, some states have built "fish ladders," water-covered steps alongside the dams that allow the fish to jump from one level to the next as they would over a natural waterfall. In spite of all this technological aid, however, the numbers of salmon have still declined, as have the numbers of other animals who depend upon the salmon for food.

or swim immediately. Many are born with fur and with their eyes and ears open. Others, however, like baby muskrats, are born hairless and blind. Mammals feed their young with milk produced by the mother.

Common river and stream mammals Mammals found in and around rivers and streams include manatees, capybaras, tapirs, minks, water shrews, beavers, dolphins, otters, and hippopotami.

FRESHWATER DOLPHIN Freshwater dolphins are found in the Ganges, Indus, Amazon, and Orinoco rivers. They are almost blind and find their way through muddy waters by sending out sound waves that bounce off obstacles to warn them of what's ahead and help them find the fish on which they feed. Like their saltwater cousins, freshwater dolphins are considered intelligent and some fishermen use them to herd schools of fish into waters where they are more easily caught.

Dolphins produce a single calf that is capable of swimming and breathing within the first minutes of its life. Some mothers have been observed guiding the calf to the surface, as if to help it reach the upper air.

RIVER OTTER River otters have webbed feet and strong tails that make them good swimmers, and their thick, water repellent fur coats keep them warm and dry. They feed on frogs, birds, fish, and small mammals. Their eyes, ears, and nostrils are placed high on their heads, which enables them to look sharply around while swimming. They dig dens into the riverbank where they care for their young until the young are able to live on their own. Otters are playful animals and spend much of their time at acrobatic games.

HIPPOPOTAMUS The name hippopotamus means "river horse" in Greek, and, although hippos look ungainly, they are fast both in the water and on land. They graze on vegetation on land at night and spend their days dozing in the muddy waters of African rivers. Adult males may reach a length of 14 feet (4 meters) and weigh as much as 4 tons (3.6 metric tons). Their skin is nearly hairless, and the eyes and nostrils protrude so that they are above the water even when the rest of their body is submerged.

A river otter eating a fish. Otters are one of the most common mammals found around rivers and streams. (Reproduced by permission of Corbis. Photograph by Mary Ann McDonald.)

ENDANGERED SPECIES

Seven types of salmon and two species of trout are on the endangered list in the northwestern United States. The Topeka shiner and the Arkansas River shiner are also threatened in the central United States and the snail darter in Tennessee. As rivers and streams become polluted, many species of birds, such as the fish eagle, are also threatened. Because birds cannot use the polluted waters for food or as nesting areas, their numbers are declining as a result.

At one time hippos were found in almost all African rivers. However, as the human population has grown, their habitat has shrunk and they are now found in only a few rivers. The Amazonian manatee, the only species of freshwater manatee, is now protected against hunters, who kill it for its meat.

HUMAN LIFE

Rivers have always been intimately connected to human life, and river valleys have been the birthplaces of the world's great civilizations. The Nile River is the scene of the oldest and perhaps the most remarkable of ancient civilizations, that of Egypt. The Nile's yearly floods brought rich sediments to the soil, and the people learned to divert the floodwaters for agriculture and drain the swamps. Chinese civilization got its start in the lower valley of the Huang He, where many floods and extreme weather forced people to develop the technology necessary for life to continue successfully there. In the valley of the Tigris and Euphrates rivers, Sumerian civilization developed as people fought to clear jungle swamps and manage floodwaters. The same was true of the Indus civilization, which arose in the valleys of the Ganges and Indus rivers in India and Pakistan. While rivers offered opportunities for water and transportation, they also presented challenges that inspired people to excel in order to survive.

IMPACT OF RIVERS AND STREAMS ON HUMAN LIFE

People use rivers and streams for water and food; boundaries; transportation and trade; recreation and building sites; and creating hydroelectric power.

Water Many rivers and streams are sources of drinking water, as well as water for such things as bathing and laundry. In developed countries, about half the freshwater supply is used by industry. In

---

STUCK IN THE MIDDLE?

The duck-billed platypus, which is found in Australian rivers, appears to be a cross between a reptile, a duck, a beaver, and an otter. It has a ducklike beak, webbed feet, and lays eggs. But it also has dense fur and a flat, beaver-like tail, and suckles its young. It is a good swimmer and dines on the river's insect population. Officially, the platypus is a mammal and cousin to the anteater, and some scientists believe that the species originated during the evolution of reptiles into mammals but never quite completed the transition.

RIVER OF DEATH

In Greek mythology, the River Styx traveled through the underworld, the land of the dead. The water of that river was supposedly so poisonous that it would dissolve any vessel used to carry it except one made from a horse's hoof. One account reports that Alexander the Great (356–323 B.C.), the king of Macedon, was poisoned by Styx water. There may be a kernel of truth in the claim, for some experts believe he died of a bacterial infection caught by drinking the water of the Euphrates River, which carried raw sewage out of the ancient city of Babylon in what is now southern Iraq.

---

less wealthy countries, 90 percent is used for irrigation (watering) of crops. The earliest irrigation systems were built on the banks of the Nile River about 6,500 years ago. Pumping systems and waterwheels soon followed. Today, about 385,000,000 acres (155,700,000 hectares) of land is irrigated worldwide.

Food The fish and plants in rivers and streams are often a source of food for humans. Although most fish used for food come from the ocean, commercially important freshwater fish include trout and salmon. In addition, farmers often use aquatic plants, such as marsh grass, reeds, and sedges, for feeding livestock.

Boundaries Throughout history, rivers have acted as boundaries. If they were too big or too fast to cross easily, they limited overland travel. During exploration of the American West, for example, wide rivers like the Missouri meant that travelers had to build a boat or raft of some kind or find a spot where the channel was narrow and the water shallower, so that horses could swim across.

Rivers have also acted as boundaries between two different territories. Many countries use rivers to help define their borders. The Rio Grande, for example, separates Texas from Mexico, and the Congo in Africa divides Congo and Zaire.

Transportation and trade Rivers have always provided a means of transportation and trade. Cities built at the edge of rivers could import or export goods upstream or downstream or even to sea more easily and less expensively than cities reachable only by overland routes. An example is the Mackenzie River of Canada's Northwest Territories, which flows into the Arctic Ocean. Early European settlements, such as Fort Good Hope, were built along its shores as part of the fur trade, and today it is used to transport petroleum and uranium ore.

Many rivers are still important routes for bulk cargo, such as coal. The Chang Jiang River in China, for example, is navigable by ocean-going freighters for 650 miles (1,046 kilometers) inland and the St. Lawrence between the United States and Canada for about 1,300 miles (2,092 kilometers).

SAVING THE SALMON

Saving threatened species of animals usually requires sacrifices on the part of humans who live in the region. When seven species of salmon that breed in the Columbia and Willamette Rivers in the northwestern United States were put on the endangered list in 1999, people living in Seattle, Washington, Portland, Oregon, and other communities agreed to:

- limit logging and move that activity back from riverbanks so as to decrease erosion

- recycle water used for golf courses and limit the use of fertilizers on golf greens

- redesign some hydroelectric dams and remove others to allow the salmon to reach their breeding grounds

- keep domestic cattle away from rivers, reduce the cattle population, and treat animal waste before it is used as a fertilizer

- limit washing of cars

- limit use of pesticides and herbicides

- build environmentally friendly sewage treatment plants

- limit salmon fishing

- enact new pollution controls for industry

Recreation and building sites Although pollution makes most of them undesirable for swimming, streams and even some rivers are popular for sport fishing and boating. In urban areas, the areas along river banks have become popular sites on which to build homes. Some rivers and streams may be diverted to create artificial lakes or ponds in order to add beauty and wildlife to a residential area.

Hydroelectric power When river waters rush over the blades of a turbine (energy-producing engine), the blades turn rapidly, and this rapid turning produces electricity. This form of electricity, called hydroelectric power, can provide heat and light for many people. Worldwide, about 6,000,000 kilowatts of energy are produced by means of hydroelectric power plants, such as that in Hoover Dam on the Colorado River in Nevada and Grand Coulee Dam on the Columbia River in Washington.

Other resources Plant materials found along rivers and streams can be used for building. Reeds are used for huts in Egypt and stilt houses in Indonesia. Also, river sediments, such as clay and mud, are used to produce bricks.

Fish are used for more than just food. They yield such products as fish oils, fish meal, fertilizers, and glue. Other animals may be introduced into rivers to help control problems. The Amazonian manatee, for example, was placed in reservoirs because they eat the river plants that were clogging the turbines of hydroelectric plants.

The water in rivers and streams may be used to cool power stations and for other industrial purposes. However, some industries have used them as dumpsites for industrial chemicals.

> ## THE "ROCK BREAKER"
>
> Henry Morton Stanley (1841–1904) was given the nickname "rock breaker" for his ability to overcome difficulties, and it was a name he certainly earned. An illegitimate child, Stanley spent part of his young life in a workhouse in Wales, the country of his birth, where he was routinely mistreated. At the age of 15, he ran away to America where a kindly merchant adopted him and turned his life in a new direction.
>
> Stanley served as a soldier in the American Civil War, as a seaman in merchant ships, and most importantly as a journalist. In 1869, he was commissioned by the *New York Herald* to travel to Africa to find David Livingstone (1813–1873), a British explorer who was missing. Stanley succeeded, and together he and Livingstone explored Lake Tanganyika in eastern Africa. After Livingstone died, Stanley carried on his explorations, eventually tracing the route of the Congo River and helping to found the Congo Free State. He described his dangerous journey in the book titled *Through the Dark Continent*, which was published in 1878.

IMPACT OF HUMAN LIFE ON RIVERS AND STREAMS

Initially, the affect of human life on rivers and streams was small. As the human population has grown, however, abuses have increased as well, and many rivers and streams resemble open sewers.

Each year, American Rivers, a conservation group, publishes a list of American rivers that are most endangered. Those endangered in 1999 were:

- Lower Snake River in Washington
- Missouri River in Montana, North Dakota, South Dakota, Nebraska, Iowa, Kansas, and Missouri
- Alabama-Coosa-Tallapoosa river system in Georgia and Alabama
- Upper San Pedro River in Arizona and Mexico
- Yellowstone River in Montana and North Dakota
- Cedar River in Washington
- Fox River in Illinois and Wisconsin
- Carmel River in California
- Coal River in West Virginia
- Bear River in Utah

Water supply Although all water on Earth is in constant circulation, some regions may have a limited supply. As populations grow, the supply diminishes even more. Forests, for example, are cut down and replaced by farms. Because trees help conserve underground water, much of this water is being lost. Because rivers and streams are often linked to groundwater, they gradually disappear, as well.

> ## THE AFRICAN QUEEN
>
> An exciting mixture of comedy, adventure, and romance, *The African Queen* is a classic film about two people during World War I (1914–18) who take an ancient, ramshackle boat up a treacherous river in Africa in hopes of finding and destroying an enemy ship. They encounter rapids, leeches, hippos, and other river dangers during the wild ride to their destination. The film starred Katharine Hepburn and Humphrey Bogart and won the Academy Award in 1951. It was based on the book of the same name by British author C. S. Forester (1899–1966).

Irrigation practices can also cause damage, especially in desert regions, because there is not enough precipitation to replace the water that is used from the rivers and streams.

Use of plants and animals The distribution of many popular species of fish has been affected by changes made to rivers and streams in Europe and America. The giant catfish, the Atlantic salmon, the Danube salmon, the sturgeon, the beluga, and many others are now found in only a few rivers. In some rivers, such as the lower Rhine in Europe, bacteria are now the predominant lifeforms.

Overdevelopment Although people may wish to live near a stream or river for its beauty and a sense of being close to nature, overdevelopment can result in erosion and loss of wildlife, as well as the scenic value. Removal of trees during housing development can cause ground water to disappear and deprive the river or stream of its source. Large tracts of homes and areas covered by concrete can prevent the rain from soaking into the ground. Instead, the runoff enters rivers and streams, causing flooding.

Quality of the environment Pollution by communities, industries, shipping, and poor farming practices has led to poisoning of water and changes in its temperature, as well as loss of animal habitats.

The most common form of river pollution is domestic sewage. If there is enough oxygen in the water, bacteria can break the sewage down quickly. Problems occur when the volume of sewage is so large and bacteria use up so much oxygen that plants and animals cannot survive. Additionally, this process results in the formation of ammonia, a gas that is poisonous to most animals. Sewage fungus, a combination of bacteria and molds, may grow on the surface of the water.

Another source of river pollution is industrial chemicals and heavy metals, which may dissolve in the water or remain as solids. Enormous quantities of these materials poison animals, clog gills in fish, and bury the riverbed in sediments. Industries are also responsible for dumping heated water (thermal pollution) into rivers, raising their temperature and killing many species of plants and animals that require a cooler environment.

GIANTS OF THE RIVER

The last part of the twentieth century has seen the building of colossal dams designed to bring hydroelectric power and flood control to undeveloped countries.

As of 1999, the world's largest was the Itaipu Dam on the Parana River between Brazil and Paraguay in South America. Completed in 1991 at a cost of over $20 billion, it has 18 huge turbines that generate as much as 12,600 megawatts of electricity. (By comparison, the largest dam in the United States, the Grand Coulee, outputs only 9,700 megawatts.) Over 30,000 workers took 16 years to create the 4.8-mile- (7.6-kilometer-) long, 500-foot- (152-meter-) high structure, using enough concrete to build eight medium-sized cities. Its reservoir covers 870 square miles (2,262 square kilometers), and thousands of farm families had to be relocated. The presence of the reservoir has caused some cooling of the local climate, which endangers wheat and other commercial crops, but the dam produces clean, renewable energy and creates thousands of jobs.

Soon to overtake the Itaipu in scope and power production is the Three Gorges Dam on the Chang Jiang (Yangtze) River of China. With its 26 turbines, the Three Gorges is expected to output 18,200 megawatts of electricity—as much as 18 nuclear plants—and help bring prosperity to central China. At a projected cost of $75 billion, the dam is scheduled for completion in 2009 and will be 607 feet (185 meters) high and 1.3 miles (2 kilometers) long. Its reservoir will be 370 miles long (592 kilometers), filling the towering limestone canyons along the river's route. As many as 14 towns will be drowned and 1,900,000 people relocated. Although some people oppose the dam because it will kill many aquatic animals and destroy valuable archeological sites, it is hoped that the dam will end the disastrous flooding common to the Chang Jiang, which has killed as many as 300,000 people during the past century alone.

Other giant dams in progress include the Ataturk Dam on the Euphrates River in Turkey and the Sardar Sarovar in India, part of a proposed 30-dam project on the Narmada River.

Shipping can also cause pollution. Ships discharge waste into the water or develop leaks and spills that threaten the water's purity and destroy wildlife.

Eutrophication (yoo-troh-fih-KAY-shun) occurs when fertilizers used in farming get into rivers and streams, spurring a greatly increased growth of algae in slow-moving waters. These plants form a thick mat on the surface and block the sunlight, causing submerged plants to die. As the dead plants decay, oxygen in the water is used up, killing fish and other plant life.

Flood control Flooding nourishes the soil in floodplains, creates natural wetlands, and supports wildlife. However, floods can also be destructive and dangerous to human life. For this reason, efforts have long been made to control floods by means of dams and artificial levees. Unfortunately, many of these changes have not only harmed wildlife, they have sometimes caused more problems than they have helped. The Aswan High Dam in Egypt is an example. Completed in 1970, the dam is 2.3 miles (3.3 kilometers) in length and rises 364 feet (111 meters) above the Nile riverbed. The dam was built in order to irrigate desert regions. However, dissolved salts have built up in the irrigated areas and increased the saltiness of the Nile itself. Formerly, the Nile carried vast amounts of sediments to the Mediterranean. With their reduction, the sardine fishery in the eastern Mediterranean has been destroyed. Also, the dam has increased erosion in the Nile waterway and in the delta region.

A group of fish killed by water pollution. (©U. S. Fish & Wildlife Service)

In Quebec, Canada, many people are opposing the James Bay Project, a proposed hydroelectric dam that will alter the region. More and more people have begun to wonder if it may be wiser to allow rivers to flow unobstructed.

Wild and Scenic Rivers Act In 1968, the U.S. Congress passed the Wild and Scenic Rivers Act, which established a program to study and protect free-flowing rivers. The act prohibits building hydroelectric or other water development projects on certain rivers. In order to qualify for protection, a river must be undammed and have at least one outstanding resource, such as a wildlife habitat or historic feature. Currently, 152 rivers fall under the Act's protection, including the Klamath River in California and the Rio Grande in Texas.

SWAN SONG FOR THE THAMES

Even before 1800, the Thames (TEMS) River in England was polluted by chemicals and other materials used in manufacturing. With the invention of flush toilets, things got much worse. Where did the sewers empty out? In the Thames. The fish suffocated to death, and the swans and other waterbirds left for a healthier part of town. Until the 1950s, the Thames was considered one of the most polluted rivers on the planet.

Finally, public protest grew so loud that things began to change. Strict new regulations were created to limit what could be poured into the water. Sewage-treatment plants were improved, and special equipment was installed in some locations to mix the water with oxygen. Before long, the cure began to take hold. Within 30 years, pollution had been reduced by 90 percent, and oxygen content rose to healthy levels. The fish returned, and so did aquatic plants. Even the swans moved back.

NATIVE PEOPLES

Rivers have always attracted human settlements, and most rivers and streams have probably played a part in the lives of people native to the region. In the United States, for example, the Great Plains between the Rocky Mountains and the Mississippi River was the home of many Native American tribes, including the Arapaho, Blackfoot, Cheyenne, Crow, Iowa, Pawnee, and Sioux. Although the buffalo were the primary source of meat, the rivers and streams provided fish and other edibles, as well as water, and villages were often located alongside them. Rafts and dugout canoes made from logs, as well as more sophisticated craft covered with skins or birchbark, were used for river travel.

Since 1940, native peoples who still live a traditional lifestyle are usually found in undeveloped countries, and most of those have been touched by the modern world in some ways. The Jivaro of Ecuador and Peru are an example. They live in the tropical forests and depend upon the rivers of that region—the Tigre, the Pastaza, the Morona, the Alto Moranon, and the Santiago—for fish and a means of travel.

THE FOOD WEB

The transfer of energy from organism to organism forms a series called a food chain. All of the possible feeding relationships that exist in a biome make up its food web. In rivers and streams, as elsewhere, the food web consists of producers, consumers, and decomposers. These three types of organisms transfer energy within the biome.

Algae are the primary producers in rivers and streams. They produce organic materials from inorganic chemicals and outside sources of energy, primarily the Sun.

Animals are consumers. Those that eat only plants, such as snails, are primary consumers in the river or stream food web. Secondary consumers, such as carp, eat the plant-eaters. Tertiary consumers are the predators, like otters and anacondas, that eat second-order consumers. Humans are omnivores and eat both plants and animals.

Decomposers, which feed on dead organic matter, include some fly larvae. Bacteria also play an important part in decomposition.

Extremely damaging to the river or stream food web is the concentration of pollutants and dangerous organisms that become trapped in sediments where life forms feed on them. These life forms are fed upon by other life forms and, at each step in the chain, the pollutant becomes more concentrated. Finally, when humans eat fish, ducks, or other river wildlife contaminated with the pollutants, they, too, are in danger of serious illness. The same is true of diseases such as cholera, hepatitis, and typhoid, which can survive and accumulate in certain aquatic animals. These diseases can then be passed on to people who eat the animals.

> ## ROADS TO ADVENTURE
> During the great period of world exploration, from about 1400 to 1900, most inland travel was accomplished by means of rivers. The journeys of Europeans into the North American wilderness are no exception. Jacques Cartier (1491–1557), a Frenchman, was the first. Beginning in 1534, he traveled down the St. Lawrence River to where Montreal, Quebec, is today. In 1673, Louis Jolliet (1645–1700) and Jacques Marquette (1637–1675), also French, traveled down the Mississippi River to where it joins the Arkansas River, then returned via the Illinois River to Lake Michigan. Alexander Mackenzie (1764–1820), a Scot, descended the Mackenzie River from Great Slave Lake in the Northwest Territories of Canada to the Arctic Ocean in 1789.

SPOTLIGHT ON RIVERS AND STREAMS

THE AMAZON

Of all the rivers in the world, the Amazon carries the greatest volume of water—one-fifth of all river water—and has the largest drainage basin—5 percent of the world's total land area. In 1994, researchers claimed to have discovered its true source as the Ucayli River, and, if they are right, that may make the Amazon also the world's longest river, surpassing even the Nile. Its discharge—up to 7,000,000 cubic feet (196,000 cubic

> ## THE AMAZON
> **Location:** Brazil and Peru in South America
> **Length:** 3,915 miles (6,437 kilometers)
> **Drainage Basin:** 2,722,000 square miles (7,077,200 square kilometers)

meters) per second—is so strong that it carries fresh water 200 miles (320 kilometers) into the Atlantic Ocean.

More than 1,000 streams make up the Amazon's tributaries, at least 20 of which are more than 600 miles long and are major rivers in their own right. They include the Ucayli, the Tocantins, the Xingu, the Tapajos, the Madeira, the Purus, the Jurua, the Javari, the Huallaga, the Maranon, the Trombetas, the Negro, the Casiquiare, the Japura, the Putumayo, the Napo, the Nanay, the Pastaza, the Morona, and the Tigre. The main stream originates in glacier-fed lakes in the Andes Mountains of Peru, only 100 miles (160 kilometers) from the Pacific Ocean, and travels across the continent to empty into the Atlantic.

Average velocity is about 1.5 miles (2.4 kilometers) per hour, although velocity increases greatly at flood times. During a flood, the height of the river may increase as much as 50 feet (15 meters). At its mouth, a 12-foot (3.6 meter) high tidal bore occasionally travels upstream at about 15 miles (24 kilometers) an hour with a force that can uproot trees growing along the river's banks.

The Amazon also receives up to 8 feet (2.4 meters) of annual rainfall, which causes it to overflow its banks for up to 30 miles (48 kilometers) on either side. This vast floodplain is underwater for several months each year, and many channels and oxbow lakes form. In most of the Brazilian parts of the river, its depth is more than 150 feet. The maximum width of its permanent bed is 8.5 miles (13.6 kilometers). The width of its mouth is 207 miles (538 kilometers). No delta has formed because the 1,500,000 tons (1,360,500 metric tons) of sediment discharged daily are carried northward by ocean currents and deposited along the coast of Guiana.

The Amazon basin supports the largest area of rain forest in the world and the greatest plant and animal diversity. About 100,000 species of plants, such as cassava, tonka beans, guava, and calabash, are found here. Trees include palms, myrtles, laurels, acacias, and rosewoods.

The Amazon region hosts more than 2,000,000 species of insects, and the number of Amazon fish species has been estimated at 2,000. A few, such as the piraucu, are commercially important. Most streams are home to piranhas, sting rays, and giant catfish.

Turtles and giant constrictor snakes represent reptiles. Birds live among the trees, and some, such as the fish eagle, hunt food in the river. Mammals include the capybara.

Most human inhabitants of the Amazon basin have been Indians, such as the Jivaro and the Yanomamo, but the area has never been heavily populated. The soil is poor and not really suitable for farming, although huge tracts of forest are being destroyed for agricultural purposes. The river itself has been used for transportation and trading since the earliest times, and ocean-going vessels can travel as far upstream as Manaus, a city west of Brazil.

For more information about the Amazon region, see the chapter titled "Rain Forest."

THE NILE

The Nile has historically been considered the longest river in the world, although in 1994 some researchers claimed the Amazon was the longest, and the question is currently unresolved. Its main tributaries are the White Nile, which has its headwaters in Burundi, and the Blue Nile, which rises in the highlands of western Ethiopia. Lake Albert, Lake Victoria, and Lake Tana all contribute water. The Blue Nile and the White Nile join together at Khartoum in Sudan and continue north into southern Egypt. Six stretches of rapids and waterfalls break its flow, beginning at Aswan. Between Aswan and the Mediterranean Sea, the river creates a wide, rich floodplain that has been cultivated for more than 3,000 years. The Nile delta lies north of Cairo, a 100-mile- (160-kilometer-) long stretch of rich silt deposits cut by many streams.

> **THE NILE**
>
> **Location:** Egypt, Burundi, Rwanda, Uganda, Ethiopia, Kenya, Tanzania, and Sudan in Africa
>
> **Length:** 4,157 miles (6,651 kilometers)
>
> **Drainage Basin:** 1,100,000 square miles (2,860,000 square kilometers)

Precipitation varies along the Nile's length, but most of its basin receives no rain from November to March. At Khartoum, 6 inches (15 centimeters) of rain fall annually, but Cairo receives only 1 inch (2.5 centimeters). The far south, however, receives 50 inches (127 centimeters) each year. Before the Aswan High Dam was completed in 1970, heavy rains in the south caused floods in Egypt. Although the floods were vital to agriculture in the region, they often caused serious damage as well. The dam now allows for some flood control.

Where the river passes through Ethiopia and eastern Africa tropical forests and grasslands occur. Papyrus, water lettuce, sedges, and water hyacinth grow in and around the river. Farther north, drier conditions prevail and support only acacia and scrub. Beyond that, the Nile passes through true desert. In Egypt, desert land along the Nile is irrigated and cultivated.

The river supports fish, snakes, turtles, crocodiles, and lizards. In its southern waters, hippopotami can be found.

About 3100 B.C., the ancient Egyptian civilization began on the banks of the Nile and was totally dependent upon the river both for water and for the sediments that enriched the soil. Lower Egypt, which was centered around the delta, was fertile and green. Upper Egypt was hot and dry, and the band of land suited for farming was only a few miles wide. Crops of all kinds were grown, however, and cattle, goats, pigs, and sheep were pastured.

The river provided fish and waterfowl, and papyrus and other reeds and grasses were turned into fibers for baskets, boats, and paper. River mud was used to make bricks and, over time, the Egyptians developed great architectural skill, creating the magnificent pyramids which still survive.

Today, almost 99 percent of Egypt's population lives in the Nile valley and delta. Fishers and farmers live in the irrigated north and in southern regions where rainfall is adequate. In the northern Sudan, nomads raise cattle and camels. In more arid (dry) regions, the economy is extremely dependent upon the river. The Aswan High Dam produces electricity and has increased irrigated land by 20 percent, although some people are concerned that the dam is decreasing the fertility of the soil and causing the water in the delta region to become more salty.

THE HUANG HE (YELLOW RIVER)

The Huang He (wang HO) or Hwang Ho, which is Chinese for Yellow River, is the second longest river in China and flows eastward from the high country of Tibet through the central plains to the Yellow Sea. Its waters are clear in the upper regions, but as it passes through Shanxi and Shanxi provinces, it picks up the large quantities of sediments—1,600,000,000 tons (1,440,000 metric tons) annually—that give it its yellow color and therefore its name. These sediments make the Huang He's delta the fastest growing in the world. Each year it extends another 1.2 miles (2 kilometers) into the sea.

> ### THE HUANG HE (YELLOW RIVER)
> **Location:** China
> **Length:** 2,901 miles (4,641 kilometers)
> **Discharge Basin:** 486,000 square miles (1,263,600 kilometers)

Average precipitation in the Huang He basin is about 16 inches (40 centimeters) a year, and average annual discharge is 11 cubic miles (48 cubic kilometers). Rainfall varies greatly, however. In winter and early summer, the river is at its low point. In spring and late summer, water levels are high, sometimes as much as 30 feet (9 meters) above its banks, and severe flooding is common. The Huang He has overflowed its dikes (retaining walls) hundreds of times in the past 3,000 years, causing so much damage and so many deaths that it has been referred to as "the sorrow of China." In an attempt to control flooding, artificial embankments have been created along 1,120 miles (1,800 kilometers) of the river's length.

Much of the Huang He basin is grassland. Cattail marshes are home to many geese, ducks, and swans. Carp and catfish live in the river itself.

China's great civilization, dating from at least as early as 2000 B.C., probably began in the Huang He basin. Today the basin supports a population of more than 100,000,000 people. In many places, the current of the Huang He is so swift and the channel so shallow that it is of little use for navigation, but the floodplain is developed for agriculture.

THE CHANG JIANG (YANGTZE)

The Chang Jiang (zhang jee-ANG), also spelled Yangtze, is the longest river in China. It crosses the central region from east to west, where it flows into the East China Sea near Shanghai. Its headwaters are in the Tibetan highlands at elevations of about 17,000 feet (5,168 meters), and its course makes many bends. Below Wan-hsien it travels for about 200 miles (320 kilometers) through spectacular gorges and deep canyons where rapids form. In the region of the gorges, the river is from 500 to 600 feet (152 to 183 meters) deep, making it the world's deepest. During summer floods, the water in this region often rises 200 feet (61 meters) and travels at velocities faster than 13 miles (21 kilometers) per hour. Boats are often towed through the gorges by large gangs of men on shore using half-mile-long (0.8-kilometer-long) ropes made from bamboo.

> **THE CHANG JIANG (YANGTZE)**
> **Location:** China
> **Length:** 3,434 miles (5,494 kilometers)
> **Drainage Basin:** 756,498 square miles (1,966,895 square kilometers)

Farther east, near Hankow, is the only bridge across the Chang Jiang. Completed in 1957, the bridge is 3,762 feet (1,144 meters) long. Here, the river forms a vast floodplain, and flooding often occurs in summer. In 1998, the worst flood in 44 years caused 3,004 deaths and $20 billion in damages. More than 14,000,000 people were left homeless.

Over its length, precipitation averages 40 inches (100 centimeters) annually. Average discharge into the sea is 700,000 cubic feet (19,600 cubic meters) per second, and 1,000,000,000 tons (907,000,000 metric tons) of sediment is deposited annually. The delta is extending into the sea at a rate of 1 mile (1.6 kilometers) every 64 years.

Deciduous forests cover those areas of the basin that are not cultivated for farming. About 70 percent of China's rice crop is grown in the Chang Jiang basin, and about 300,000,000 people live in the region.

The giant paddlefish lives in the Chang Jiang, as does small species of alligator. Animal life varies along its length and includes thrushes, pheasants, and antelopes.

The Chang Jiang is important commercially, and the volume of transportation is greater than that of all the other waterways of China combined. About 25,000 miles (40,000 kilometers) of main river and tributaries can be navigated by small craft, and ocean-going ships can reach Wuhan, a city in central China. Five of China's largest cities—Shanghai, Wuhan, Chongqing, Chengdu, and Nanjin—are on or near the Chang Jiang system.

The river is considered ideal for generating hydroelectric power, and the gigantic Three Gorges Dam, which was begun in 1993, is predicted to provide at least 15 percent of China's electricity needs. The dam will also help control flooding, although many scenic areas will be permanently covered by its reservoir. Thousands of people are being forced to relocate, and some scientists believe that the dam will have undesirable affects on the environment after it is completed, around 2009. (See sidebar, p. 378)

THE GANGES (GANGA)

The Ganges (GANN-jeez) crosses northern India and then flows south through the provinces of Uttar Pradesh, Bihar, West Bengal, and finally Bangladesh. Its headwaters are the Alaknanda and Bagirathi Rivers, which begin in the Himalaya Mountains. Its delta begins 200 miles (320 kilometers) from the sea, where it joins with the Jumna River to form the Padma and then flows into the Bay of Bengal.

Although the flow of water in the Ganges varies with the seasons, the changes are seldom violent. It travels through dry plains, drawing water from several tributaries such as the Jumna, the Gogra, the Gandak, the Kasi, and the Brahmaputra. Its flow is sluggish, although floods are common, especially in the delta region where tropical storms and tidal waves (giant waves associated with earthquakes) occur. During one such storm in 1876, about 100,000 people were drowned in half an hour.

> ### THE GANGES (GANGA)
> **Location:** India and Bangladesh
> **Length:** 1,557 miles (2,491 kilometers)
> **Discharge Basin:** 188,800 square miles (490,880 square kilometers)

Forests grow on the Gangetic plain. Where rainfall is heavy, evergreens predominate. Carp are found in the river itself as are geese and ducks. Mammals include tigers, deer, and wild dogs, but reptiles are more numerous, especially crocodiles, lizards, and snakes.

Changing water levels limit the use of the Ganges for large vessels. In the delta region, many places are accessible only by small water craft. Soils here are rich and fertile, but in the north the land is too high for irrigation unless power equipment is used. Crops include grains, sugar, cotton, and lac trees.

To those of the Hindu faith, the Ganges is holy, and many places along its shore, such as Varanasi (Benares), Allahabad, and Hardwar, have religious significance. Many people go on pilgrimages (religious journeys) to visit these places. Hindus believe that bathing in the Ganges is religiously purifying, and thousands flock to its banks each day. However, the river is polluted by human sewage, fertilizer runoff, and industrial wastes. In 1985, efforts were begun to clean up the river and are ongoing.

One of the world's earliest civilizations, the Indus, began in the neighboring Indus River valley and spread into the Ganges valley. One of the most well-known archeological sites is Mohenjo-Daro, beside the Indus River.

THE INDUS

The Indus (INN-duhs) River begins in Tibet, travels through part of northern India, and completes its course in Pakistan. Its headwaters are formed by meltwater in the Himalaya Mountains and the Karakoram Range, where its flow is turbulent, unnavigable, and prone to flooding, especially in the summer months. When it leaves the mountains and enters the dry Punjab plains of Pakistan, it broadens and picks up sediments.

> ## THE INDUS
>
> **Location:** Tibet, India, and Pakistan
>
> **Length:** 1,800 miles (2,880 kilometers)
>
> **Drainage Basin:** 372,000 square miles (967,200 square kilometers)

The major tributaries of the Indus are the Punjab, the Sutlej, the Chenab, the Jhelum, the Ravi, and the Beas. Its delta begins at Tatta and enters the Arabian Sea 70 miles (110 kilometers) farther south, where the river splits into several channels.

Rainfall in the Indus Valley ranges from 5 to 20 inches (13 to 51 centimeters) annually, and in the drier regions the Indus is used for irrigation.

Desert vegetation predominates and includes thorn forests and many shrubs. Grasses do not thrive here. Catla are common fish. Sarus cranes and bearded vultures frequent the area, as do crocodiles and cobras. Mammals include deer and wolves.

The Punjab plain is Pakistan's richest farming region, where wheat, corn, rice, millet, dates, and other fruits are grown. Large dams have been built on the river to provide hydroelectric power and water for irrigation. Use of the Indus has been disputed by India and Pakistan as they have tried to determine their borders.

The cities of the great Indus civilization, Harappa and Mohenjo-Daro, were built in the Indus valley around 2500 B.C. Although the civilization appears to have declined gradually, it is believed that barbarian nomads of the Eurasian high plains invaded the region and destroyed what was left of this civilization.

THE COLORADO

The Colorado River is created by rain and melting snow in the mountains of central Colorado, Wyoming, and Utah. Its main tributaries are the Gunnison, the Dolores, the Green, the San Juan, the Little Colorado, the Gila, and the Virgin. As it travels south and southwest through Utah, Arizona, and Nevada, it creates the magnificent Grand Canyon, cutting deeper into the rock each year. The canyon is 218 miles (349 kilometers)

> ## THE COLORADO
>
> **Location:** Colorado, Arizona, Utah, Nevada, and California
>
> **Length:** 1,440 miles (2,304 kilometers)
>
> **Drainage Basin:** 244,000 square miles (634,400 square kilometers)

long and, in one spot, 15 miles (24 kilometers) wide and 6,000 feet (1,824 meters) deep. At its mouth, the Colorado empties into the Gulf of California on the Pacific coast.

Discharge rates vary from 3,000 cubic feet (84 cubic meters) per second when water levels are low to 200,000 cubic feet (5,600 cubic meters) per second at flood times.

Much of the region is treeless and desert-like, and plants include those that tolerate semiarid conditions. Desert birds and waterfowl are common. Mammals include deer and bears.

The Colorado is one of the most valuable rivers in the world in terms of irrigation and power. Although the river is prone to flooding, Hoover Dam and its reservoir (storage area), Lake Mead, have helped prevent disasters. About 741,000,000,000 cubic feet (21,000,000,000 cubic meters) of water are available annually for use in power generation, irrigation, recreation, and community water supplies. The Los Angeles-San Diego region receives much of its water from the Colorado, and arguments have ensued among neighboring states as to how the water should be divided. Environmental concerns have slowed further development of river resources.

Until the Spanish arrived in 1540 and gave the Colorado its present name, which means "red color," the Colorado basin was the home of several Native American tribes, including the Pueblo and the Navajo.

THE CONGO (ZAIRE)

The Congo, or Zaire (zy-EER), River is the second longest in Africa. During the fifteenth century, Europeans called it the Zaire, a corruption of a Bantu word meaning "river." Later, the name was changed to Congo after the great African kingdom of Kongo located near its mouth. Today it is known by both names.

> ## THE CONGO (ZAIRE)
> **Location:** Zambia, Zaire, Congo, and Angola
> **Length:** 2,716 miles (4,346 kilometers)
> **Discharge Basin:** 1,425,000 square miles (3,705,000 square kilometers)

The Congo's headwaters are in Zambia. It then travels north across the country of Zaire, turns west, and then angles southwest, forming the border between Zaire and Congo. Its mouth, where it empties into the Atlantic Ocean, is in Angola.

The Congo's discharge basin is the second largest in the world, for it draws water from tributaries in the Central African Republic, Cameroon, Congo, Angola, Zaire, and northern Zambia. Some of its tributaries include the Chambezi, the Lualaba, the Luvua, the Lukuga, the Chenal, the Ubangi, the Sangha, and the Kwa. Along its course it forms many lakes, such as Bangweulu, Mweru, and Malebo Pool. Rapids, waterfalls, and gorges, as well as a narrow channel, are common along its length.

Because it straddles the equator, passing through the great rain forest of Zaire, it is well supplied with rainfall all year long. In the central region, annual rainfall averages 66 inches (168 centimeters). In the south the average drops to 47 inches (120 centimeters) and in the north to only 8 to 16 inches (20 to 40 centimeters).

The Congo's average rate of discharge is more than 3,000 cubic feet (84 cubic meters) per second. Its force is so great that brownish-colored Congo waters can still be distinguished for more than 300 miles (480 kilometers) out to sea. During prehistoric times its flow cut a submarine canyon as deep as 4,000 feet (1,220 meters) into what is now the ocean floor.

In the northern and southern parts of its basin it passes through grasslands (savannas) where trees are scattered. In Zaire and along the Zaire-Congo border, dense rain forest occupies the basin. During periods of flooding, the river may overflow for 6 miles (9.6 kilometers) into the forests on either side.

Vegetation in the central basin is abundant and diverse. Huge rain forest trees draped with clinging vines line its banks. In the southern regions, acacias are common. In the highlands, bamboo, heather, and giant senecio are found. However, cultivation and burning have altered the vegetation along much of the Congo's length, and many non-native plants, such as cassava, citrus, and maize have been introduced.

More than 1,000 species of fish have been identified in Congo waters. Among the best known is the lungfish. Mammals typical of grasslands, such as lions and zebras, are found in the northern and southern regions, while gorillas and rhinoceroses are found in the rain forests. Elephants and leopards are common throughout.

Most native peoples who live along the Congo catch its fish. Some are also farmers. Many peoples, such as the Bobangi and the Teke, use the river for trade. The river is also important to Zaire as a source of hydroelectric power. However, the country's economy has not yet been able to take full advantage of this potential.

In the late nineteenth century, many European explorers, such as David Livingstone (1813–1873) and Henry Morton Stanley (1841–1904), penetrated the African interior by traveling on the Congo. (See sidebar, p. 376.)

For more information about the Zaire rain forest, see the chapter titled "Rain Forest."

THE TIGRIS AND EUPHRATES

The Tigris (TY-grihs) river flows through southeastern Turkey and Iraq. Above Basra in southern Iraq, it joins the Euphrates (yoo-FRAY-teez) to form the Shatt-al-Arab, which empties into the Persian Gulf.

The Tigris-Euphrates basin is a hot, dry plain with daytime temperatures as high as 120°F (49°C). Average annual rainfall is only 8 inches (20 centimeters), and it falls seasonally, from November to March. The Euphrates floods twice during the year, the largest occurring in April and May, and creates marshy regions along its banks.

Desert vegetation, such as thorn and scrub, predominate. Bats, jackals, and wildcats are common mammals, but reptiles are more numerous, particularly snakes and lizards. Many fish live in the river, including the Tigris salmon.

Both rivers are used for irrigation, and two major reservoirs are located on the Euphrates where farmers raise cereal grains and dates and nomadic tribes raise livestock. The Tigris provides irrigation waters for the cultivation of wheat, barley, millet, and rice.

> ## THE TIGRIS AND EUPHRATES
>
> **Location:** Turkey, Syria, and Iraq
>
> **Length:** Tigris 1,181 miles (1,890 kilometers); Euphrates 1,740 miles (2,784 kilometers)
>
> **Discharge Basin:** Tigris 145,000 square miles (377,000 square kilometers); Euphrates 295,000 square miles (767,000 square kilometers)

The Tigris is navigable to small boats only between Baghdad and al-Qurnah. The Euphrates is navigable only by native craft. Bridges across both rivers were damaged by bombing in 1991 during the Persian Gulf War.

The Tigris-Euphrates basin was the birthplace of the great Sumerian civilization, which dates back to at least 3000 B.C. in what was then Mesopotamia. The Sumerians dug canals along the Tigris, and Nineveh, the capital of ancient Assyria, was located there.

THE MISSISSIPPI

The Mississippi is the second longest river in the United States, and its name in the Algonquian language means "father of waters." Its headwaters are in Minnesota, and its drainage basin includes at least part of 31 states and covers about 40 percent of the country. As it flows south, it forms the boundary between Minnesota, Iowa, Missouri, Arkansas, and Louisiana on the west and Wisconsin, Illinois, Kentucky, Tennessee, and Mississippi on the east.

> ## THE MISSISSIPPI
>
> **Location:** Minnesota, Iowa, Missouri, Arkansas, Louisiana, Wisconsin, Illinois, Kentucky, Tennessee, and Mississippi
>
> **Length:** 2,348 miles (3,757 kilometers)
>
> **Drainage Basin:** 1,243,700 square miles (3,233,620 square kilometers)

Fed by several glacial lakes, the upper Mississippi forms rapids where it joins the Minnesota River. Further south it is lined by high bluffs, and below Cairo, Illinois, it becomes an old stream, meandering through flat floodplains between natural levees.

Its most important tributaries are the Illinois, the Chippewa, the Black, the Wisconsin, the Saint Croix, the Iowa, the Des Moines, the Ohio, the Arkansas, the Red, and the Missouri Rivers. The Missouri joins the Mississippi at St. Louis and contributes 20 percent of its total discharge. The mouth of the Mississippi is at the Gulf of Mexico where its delta covers a region about 10,100 square miles (26,150 square kilometers) in area, and about 500,000,000 tons (453,500,000 metric tons) of sediments are deposited each year.

The river travels the length of the United States and precipitation varies according to region. The north receives from 20 to 40 inches (51 to 102 centimeters) annually and the south from 60 to 70 inches (152 to 178 centimeters). The southern section of the river is prone to severe flooding, and, in 1927, it rose 57 feet (17 meters) at Cairo, Illinois. Artificial levees have been built since that time to try to contain floodwaters.

Willow, oak, and pine trees are common, as are many species of grasses and aquatic plants. Fish include catfish and paddlefish. Cranes, ducks, and other waterfowl frequent the river, as do turtles, muskrats, and deer.

Ocean-going ships can travel upstream to Baton Rouge, Louisiana. Smaller ships can travel through 15,000 miles (24,150 kilometers) of the entire river system, and use of the river for transportation is increasing. Cargoes consist mainly of Midwestern grain and petroleum products from the Gulf of Mexico.

Native peoples who once lived along the river were the Ojibwa, the Winnebago, the Fox, the Sauk, the Choctaw, the Chickasaw, the Natchez, and the Alabama. The first European to reach the river inland was Hernando de Soto (c. 1496–1542) in 1541. In 1673, Jacques Marquette (1637–1675) and Louis Jolliet (1645–1700) explored the northern regions. The river system enabled Europeans to settle the central United States, and it was an important transportation route during the Civil War.

THE MISSOURI

The Missouri is the longest river in the United States and the largest tributary of the Mississippi. Its headwaters are the Jefferson, Gallatin, and Madison Rivers in Montana, and it travels through North and South Dakota to finally join the Mississippi at St. Louis, Missouri. Its own tributaries include the Little Missouri, the Cheyenne, the White, the Niobrara, the James, the Big Sioux, the Platte, and the Kansas rivers. Its drainage basin includes parts of Canada.

> ### THE MISSOURI
> **Location:** Montana, North Dakota, South Dakota, Nebraska, Iowa, and Missouri
> **Length:** 2,466 miles (3,946 kilometers)
> **Discharge Basin:** 529,000 square miles (1,375,400 square kilometers)

In Montana and North and South Dakota, it passes through high plains where the soil is poor. Farther south it enters the humid grain belt and the drier grasslands where cattle graze. The soil is rich along is southern stretches, although erosion is heavy. The river carries so much sediment that its nickname is "Big Muddy."

Precipitation ranges from 20 to 40 inches (51 to 102 centimeters) annually. The river's average discharge is about 64,000 cubic feet (1,810 cubic meters) per second. Flood season is from April to June, and the federal government has built dams and reservoirs to help control flooding, provide irrigation, and make the river navigable.

Natural vegetation in the Missouri basin consists primarily of grasses. Fish include bass and trout. Mammals such as deer and coyotes can be found in the region.

Native peoples who once lived in the region include the Cheyenne, the Crow, the Mandan, the Pawnee, and the Sioux. French explorers Jacques Marquette (1637–1675) and Louis Jolliet (1645–1700) explored the region in 1673. By 1843 farmers had begun to settle the Missouri valley. Today, the Missouri is important to traffic in bulk freight such as grain, coal, steel, petroleum, and cement.

THE VOLGA

The Volga is the longest river in Europe. Its headwaters are in the hills north of Moscow, Russia, and it empties into the Caspian Sea, which lies on Russia's southern border. It has more than 200 tributaries, the most important of which is the Kama. The Volga delta covers more than 7,330 square miles (19,000 square kilometers).

THE VOLGA

Location: Russia

Length: 2,292 miles (3,667 kilometers)

Discharge Basin: 532,800 square miles (1,385,280 square kilometers)

Melting snow and summer rains are the Volga's main source of water. Annual precipitation averages 25 inches (64 centimeters) in the north and 12 inches (31 centimeters) in the south. Spring flooding is controlled by several dams and reservoirs. The discharge rate is about 271,186 cubic feet (7,680 cubic meters) per second.

In the north the Volga passes through forested land and forms several lakes. Below that it enters a flat, swampy basin bordered by low hills. Its southern section curves around mountains and finally enters a broad floodplain.

Willow, pine, and birch are common trees. River fish include sturgeon, perch, pike, and carp. Mammals include deer, beavers, and foxes.

Although almost all of the Volga is navigable, ice closes it to traffic for as long as four months in the winter. The river carries about 60 percent of Russia's river freight, including timber, petroleum, coal, salt, farm equipment, construction materials, fish, and fertilizers.

FOR MORE INFORMATION

BOOKS

Ganeri, Anita. *Rivers, Ponds, and Lakes*. Ecology Watch. New York: Dillon Press, 1991.

Hecht, Jeff. *Vanishing Life*. New York: Charles Scribner & Sons, 1993.

Lourie, Peter. *Hudson River: An Adventure from the Mountains to the Sea*. Honesdale, PA: Boyds Mills Press, 1992.

Parker, Steve. *Pond and River*. Eyewitness Books. New York: Alfred A. Knopf, 1988.

Quinn, John R. *Wildlife Survival*. New York: Tab Books, 1994.

Sayre, April Pulley. *River and Stream*. New York: Twenty-first Century Books, 1995.

Scott, Michael. *The Young Oxford Book of Ecology*. New York: Oxford University Press, 1995.

Stidworthy, John. *Ponds and Streams*. Nature Club. Mahwah, NJ: Troll Associates, 1990.

Yates, Steve. *Adopting a Stream: A Northwest Handbook*. Seattle, WA: University of Washington Press, 1989.

ORGANIZATIONS

American Rivers
  1025 Vermont Avenue NW, Suite 720
  Washington, DC 20005
  Phone: 800-296-6900; Fax: 202-347-9240
  Internet: http://www.amrivers.org

Center for Environmental Education
  1725 De Sales Street NW, Suite 500
  Washington, DC 20036

Environmental Defense Fund
  257 Park Ave. South
  New York, NY 10010
  Phone: 800-684-3322; Fax: 212-505-2375
  Internet: http://www.edf.org

Environmental Network
  4618 Henry Street
  Pittsburgh, PA 15213
  Internet: http://www.envirolink.org

Environmental Protection Agency
  401 M Street, SW
  Washington, DC 20460
  Phone: 202-260-2090
  Internet: http://www.epa.gov

Freshwater Foundation
  2500 Shadywood Rd.
  Excelsior, MN 55331
  Phone: 612-471-9773; Fax: 612-471-7685

Friends of the Earth
  1025 Vermont Ave. NW, Ste. 300

Washington, DC 20003
Phone: 202-783-7400; Fax: 202-783-0444

Global Rivers Environmental Education Network
721 E. Huron Street
Ann Arbor, MI 48104

Greenpeace USA
1436 U Street NW
Washington, DC 20009
Phone: 202-462-1177; Fax: 202-462-4507
Internet: http://www.greenpeaceusa.org

Isaak Walton League of America
SOS Program
1401 Wilson Boulevard, Level B
Arlington, VA 22209
Phone: 703-528-1818

National Project Wet
Culbertson Hall, Montana State University
Bozeman, MT 59717

River Network
PO Box 8787
Portland, OR 97207
Phone: 800-423-6747; Fax: 503-241-9256
Internet: http://www.rivernetwork.org/rivernet

Sierra Club
85 2nd Street, 2nd fl.
San Francisco, CA 94105
Phone: 415-977-5500; Fax: 415-977-5799
Internet: http://www.sierraclub.org

U.S. Fish and Wildlife Service Publication Unit
1717 H. Street, NW, Room 148
Washington, DC 20240

World Wildlife Fund
1250 24th Street NW
Washington, DC 20037
Phone: 202-293-4800; Fax: 202-229-9211
Internet: http://www.wwf.org

WEBSITES

Note: Website addresses are frequently subject to change.

National Geographic Magazine: http://www.nationalgeographic.com

National Park Service: http://www.nps.gov

Nature Conservancy: http://www.tnc.org

Scientific American Magazine: http://www.scientificamerican.com

BIBLIOGRAPHY

International Book of the Forest. New York: Simon and Schuster, 1981.

Knapp, Brian. *River.* Land Shapes. Danbury, CT: Grolier Educational Corporation, 1992.

Pringle, Laurence. *Rivers and Lakes.* Planet Earth. Alexandria, VA: Time-Life Books, 1985.

Pritchett, Robert. *River Life.* Secaucus, NJ: Chartwell Books, Inc., 1979.

Renault, Mary. *The Nature of Alexander.* New York: Pantheon Books, 1975.

"River and Stream." *Grolier Multimedia Encyclopedia.* Danbury, CT: Grolier, Inc., 1995.

"River." *Encyclopaedia Britannica.* Chicago: Encyclopaedia Britannica, Inc., 1993.

"Saving the Salmon." *Time* (March 29, 1999): 60-61.

Zich, Arthur. "Before the Flood: China's Three Gorges." *National Geographic.* (September, 1997): 2-33.

SEASHORE

The seashore—also called the coastline, shoreline, or beach—is that portion of a continent or island where the land and sea meet. The seashore includes the area covered by water during high tide and exposed to the air during low tide, the area splashed by waves but never under water, and the area just beyond the shore that is always under water. (The tides are the rhythmic rising and falling of the sea.) Seashores vary greatly in appearance, from flat and sandy and washed by gentle waves to storm-battered and rocky or bounded by tall cliffs.

HOW THE SEASHORE IS FORMED

Seashores were first created when the continents and islands of the Earth were formed. Since then many changes have occurred. Some happened during prehistoric times, and others are still taking place.

MOVEMENT OF THE EARTH'S CRUST

The underlying structure of the shoreline depends upon the shape of the land where it meets the ocean and the type of rock of which it is a part. Earthquakes and volcanoes during prehistoric times may have helped form many shorelines. An earthquake, for example, caused part of the California shoreline to sink. The sunken area became what is now San Francisco Bay and a new shoreline was created. In regions like northern California, where earthquakes and volcanoes still occur, the shoreline may undergo many more changes in the future.

GLACIERS

During prehistoric times, glaciers (giant, slow-moving rivers of ice) may also have altered the shape of the seashore by cutting into it and leaving deep valleys behind when they retreated. Glaciers created the fjords (fee-

OHRDS; long, narrow arms of the ocean stretching inland) of Scandinavia, Greenland, Alaska, British Columbia, and New Zealand. They also created the U-shaped valleys along the coastline of southern Chile. In polar regions, glaciers are still at work, carving deep channels as they inch toward the sea.

The presence of a glacier also weighs the land down, causing it to sink. About 10,000 years ago when many glaciers receded, some coastlines rose up, and areas once under water were now above it. In some places, such as Scandinavia, the coastlines are still rising as much as 0.8 inch (20 millimeters) a year although the glaciers have been gone for millennia.

CHANGES IN SEA LEVEL

Sea level refers to the average height of the sea when it is halfway between high and low tides. Sea level changes over time. For example, when the Ice Ages ended and glaciers began to melt into the ocean, the level of the water rose. Some areas that were once exposed to the air were now covered permanently by water and new shorelines were created.

Sea level is still changing. For the past 100 years, it has risen about 0.06 inch (1.5 millimeters) per year. During 1996 and 1997, possibly because of warmer temperatures worldwide, it rose about 0.12 inch (3 millimeters). As the water creeps higher, more of the existing shoreline becomes submerged. Low-lying coastal areas in Texas and Louisiana have already been flooded.

LOCKED IN THE ICE

Exploration of remote areas, such as Antarctica, demanded people with courage and endurance. No one had more of those valuable characteristics than Ernest Shackleton (1874–1922), a British explorer who made several expeditions to that southernmost continent in the early 1900s. His most remarkable, and almost fatal, journey was made in 1914 with 27 other men on board a small ship called *The Endurance*—a name that would prove prophetic.

The Endurance became trapped in drifting ice in the Weddell Sea (part of the Antarctic Ocean) during one of the coldest winters in memory. The explorers had come prepared to spend the winter, if necessary, but the weather did not get warm enough the fol-lowing spring to free the ship. The ice crushed its forward section, and the ship sank in November of 1915. By then the men had begun to kill penguins and their own sled dogs for food.

They had managed to save three small boats and, for seven days, used them to travel to an uninhabited island. All were starved and exhausted, but taking no time to rest on the island, Shackleton and five of the strongest men set out again in one of the boats to seek aid. They finally found it on South Georgia Island farther north. Then Shackleton turned around and went back to rescue the others. Although the first three tries were unsuccessful, the fourth succeeded. Because of Shackleton's courage and endurance, everyone survived.

OFFSHORE BARRIERS

The presence of a barrier island, reef (an underwater wall made of rocks, sand, or coral), or other offshore landmass running parallel to the shore may also affect a seashore's formation. It does this by reducing the effects of wind and water. Because the force of the waves is lessened, deposits of sediment (particles of matter) are allowed to accumulate.

Coral reefs are created in tropical regions by small, soft, jelly-like animals called corals. Corals attach themselves to hard surfaces and build a shell-like external skeleton. Many corals live together in colonies, the younger building their skeletons next to or on top of older skeletons. Gradually, over hundreds, thousands, or millions of years, a reef of these skeletons is formed. Because reefs slow the movement of the water, sediments sometimes lodge in the reef allowing plants to take root. As the plants die and decay, a layer of soil is created. Eventually, the shoreline may extend, and even trees may grow there.

THE BORDERING SEA

The oceans are constantly, restlessly moving. This movement takes place in the water column (the water in the ocean exclusive of the sea bed or other landforms) in the form of tides, waves, and currents, all of which affect the shoreline. (For a more complete discussion of the ocean's water column, see the chapter titled "Ocean.")

The presence of a barrier island, such as this, may affect the formation of a seashore. (Reproduced by permission of Corbis. Photograph by Annie Griffiths Belt.)

TIDES

Tides are rhythmic movements of the oceans that cause a change in the surface level of the water. When the water level rises, it is called high tide. When it falls, it is called low tide. Along some shorelines, the tides are barely measurable. In other areas, however, the difference between high and low tide may be at least several feet (meters). High and low tides occur in a particular place at least once during each period of 24 hours and 51 minutes.

Tides are caused by a combination of the gravitational pull of the Sun and Moon and the Earth's rotation. The Sun or Moon pulls on the water, causing it to bulge outward. At the same time, the centrifugal force (movement from the center) created by the Earth's rotation causes another bulge on the opposite side of the Earth. Both of these areas then experience high tides.

THE BEACHES AT NORMANDY

Beaches have often been used by invading armies to conquer the neighboring region. During World War II (1939–45), the beaches of the province of Normandy, France, were the site of an important strike against Nazi Germany. Germany had conquered France earlier in the war, and much of their power in that country was concentrated in Normandy, which bordered the English channel and was only a short distance from England, which the Nazis also hoped to conquer.

Then on D-Day, June 6, 1944, in an effort to recapture France, the Allies (Britain, Canada, Poland, France, and the United States) struck the beaches at Normandy. The beaches, named Utah, Omaha, Gold, Sword, and Juno, stretch for about 50 miles along the coast between Cherbourg and LeHavre. The slope of the beaches is gentle and a wide expanse of shore is exposed at low tide. The date for the invasion was selected based on the tides, the weather, the presence of moonlight, and other conditions.

The Germans, who expected trouble, were unaware of the exact invasion point and had 50 infantry and 10 panzer (tank) divisions spread out over France and neighboring countries. To distract the Germans, for two months before D-Day British-based aircraft bombed rail lines, bridges and air fields on French soil. The night before, paratroops were dropped inland to interfere with enemy communications. Naval guns pounded German gun nests on shore.

Then, in the early daylight at low tide on June 6, in rough seas, about 5,000 Allied ships approached the Normandy coastline. The British and Canadians moved in efficiently at Gold, Juno, and Sword beaches and the Americans at Utah Beach. But at Omaha Beach, the central point of the landings, American troops encountered heavy German gunfire, and many men were killed. Within 5 days, 16 Allied divisions had landed in Normandy. By August, Paris had been freed, and it was the beginning of the end of Nazi rule in Europe.

A realistic account of the Normandy invasion can be viewed in Stephen Spielberg's 1998 Academy-Award-winning film *Saving Private Ryan.*

At the same time, water is pulled from the areas in between, and those areas experience low tides.

When the Earth, Sun, and Moon are lined up, the gravitational pull is strongest. At these times, high tides are higher and low tides are lower than normal. These are called spring tides. When the Earth, Sun, and Moon form a right angle and the gravitational pull is weakest, high tides are lower and low tides are higher than normal. These are called neap tides.

Tidal bores are surges of tidal waters caused when ridges of sand block the flow of ocean water and direct it into a narrow channel, sometimes as a single wave. Most tidal bores are harmless, but the bore that enters the Tsientang River in China sometimes reaches 25 feet (7.6 meters) in height and can be dangerous.

WAVES

Waves are rhythmic rising and falling movements in the water. Most surface waves are caused mainly by wind. Their size is due to the speed of the wind, the length of time it has been blowing, and the distance over which it has traveled. A breaker is a wave that collapses on a shoreline in a mass of foam. As it rolls in from the ocean, the bottom of the wave is slowed by friction as it drags along the sea floor. The top then outruns it and topples over, landing on the shore.

Waves can be extremely powerful. Storm waves can even hurl large rocks high into the air. Because so many rocks have broken the beacon at the lighthouse at Tillamook Rock, Oregon, the beacon—139 feet (42 meters) above the water—is now enclosed in a steel grating.

One dangerous type of wave, called a tsunami (soo-NAH-mee), is caused mainly by undersea earthquakes. When the ocean floor moves during the quake, its vibrations create a powerful wave that travels to the surface. When tsunamis strike coastal areas, they can destroy entire towns and kill many people.

CURRENTS

Currents are the flow of the water in a certain direction. Most currents are caused by the wind, the rotation of the Earth, and the position of continental landmasses. A longshore current is one that runs along a shoreline. Rip

THE SILVER DRAGON

High tides arrive at most shorelines gradually. In locations where tapering coastlines face a large ocean, however, the volume of the tide is focused very narrowly and its force increases. When such a tide meets a river flowing into the sea, a steep wave forms, forcing itself up the river and far inland. This wave is called a tidal bore.

The world's largest tidal bore is called the "Silver Dragon," and it occurs once a month where the South China Sea invades the Quiantang River of China. Reaching heights of 20 feet (6 meters) or more, the Silver Dragon rushes over an area 5 miles (8 kilometers) wide.

Other large bores occur on the Amazon River in South America, on the Petitcodiac where it meets the Bay of Fundy in Canada, on the Hooghly in India where it flows into the Bay of Bengal, and on the Severn in England.

currents, or riptides, are strong, dangerous currents that occur when currents moving toward shore are deflected away from it through a narrow channel.

Upward and downward movement also occurs in the ocean. Vertical currents are primarily the result of differences in the temperature and salinity (level of salts) of the water. In some coastal areas, strong wind-driven currents carry warm surface water away. Then an upwelling (rising) of cold water from the deep ocean occurs to fill the space. This is more common along the western sides of continents. These upwellings bring many nutrients from the ocean floor to the surface, encouraging a wide variety of marine (ocean) life.

IN THE PATH OF THE GREAT WAVES

Over 2.8 million people have been killed by tsunamis (giant waves) during the past 20 years. One in every three tsunamis occurs off the coast of Chile, and that country experiences 40 percent of all the damage caused worldwide.

In 1964, an earthquake struck Alaska, causing great damage and breaking a pipeline carrying oil. The oil caught fire. Then a tsunami three stories tall followed the quake and, picking up the burning oil, carried it overland in a tide of fire. The fires reached the railroad yards, where the steel tracks soon glowed red from the heat. When more tsunamis followed and flowed over the tracks, the sudden cooling made the tracks rise up and curl like snakes.

ZONES IN THE SEASHORE

The seashore can be divided into zones based on relationships to the ocean, particularly the tides.

The seashore's lowest zone is underwater at all times, even during low tide. This lower zone is also called the sublittoral zone. It is a marine environment.

The intertidal zone (also called the middle, or the littoral, zone) is covered during high tide and exposed during low tide. This is the harshest of all seashore environments, since any animal or plant that lives here must be able to tolerate being submerged for part of the day and exposed to the air and Sun for the rest.

The upper zone, which is never underwater, may be frequently sprayed by breaking waves. It is often referred to as the "splash" or supralittoral zone. Conditions here are very similar to other dry-land areas.

CLIMATE

The climate of a particular seashore depends upon its location. In general, if the shoreline is part of a desert country, such as Saudi Arabia, the climate will be hot and dry. If it is part of a country in the frozen north, such as Siberia, the climate will be cold. The presence of the ocean, however, can create some climatic differences.

The ocean absorbs and retains some of the Sun's heat. In the winter, the warmer water releases heat into the colder atmosphere, helping to keep wintry temperatures warmer inland. In summer, when the water temperature is cooler than the air temperature, winds off the ocean help cool coastal areas.

Other effects of the ocean on climate result primarily from moderating ocean currents and storms at sea.

MODERATING CURRENTS

Ocean currents may be warm or cold, depending upon where they started from. The presence of a current can moderate the weather along a particular coastline. For example, the North Atlantic Drift is a warm current that originates in regions farther south. As it flows around the coast of Scotland, it warms the air temperature enough that palm trees will grow there. As it travels farther north along the coast of Norway, it keeps the water above freezing so that Norway's ports are open all winter, even though they are above the Arctic Circle.

STORMS AT SEA

Seashores are vulnerable to storms that originate at sea and produce strong winds and high waves. Hurricanes and typhoons, for example, are

An illustration showing the upper, middle, and lower zones of the seashore.

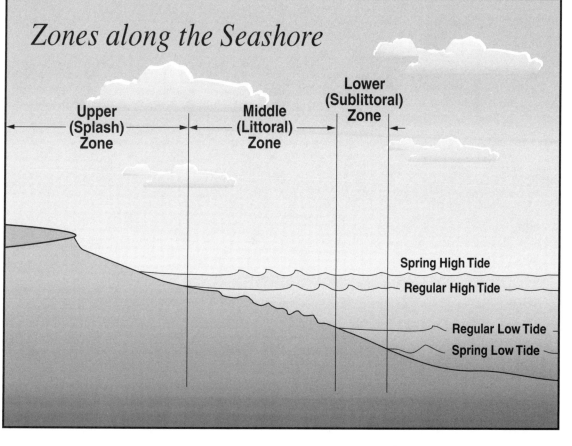

Zones along the Seashore

Upper (Splash) Zone
Middle (Littoral) Zone
Lower (Sublittoral) Zone

Spring High Tide
Regular High Tide
Regular Low Tide
Spring Low Tide

violent tropical storms that form over the oceans. Their wind speeds can reach from 75 to 180 miles (120 to 300 kilometers) per hour, and they may release as much as 10 inches (250 millimeters) of rain a day. Their forceful, rotating winds cause much damage when they reach land, as do the waves that batter the shoreline.

The worst coastal weather in the world, however, is in the North Atlantic, where the climate is cooler and where the force of the waves has been measured at 6,000 pounds per square foot (13,200 kilograms per 0.09 square meter). Driven by gale-force winds, waves along these coastlines have been known to destroy entire lighthouses as tall as 114 feet (35 meters).

GEOGRAPHY OF THE SEASHORE

The geography of the seashore is affected by the process of erosion (wearing away) and deposition (dep-oh-ZIH-shun; setting down), which helps determine the different types of surface found at the seashore, as well as its landforms.

Hurricane winds blowing palm trees off the Florida coast. (Reproduced by permission of The Library of Congress.)

EROSION AND DEPOSITION

As waves crash against a shoreline, they compress (squeeze) the air trapped in cracks in rocks. As the waves retreat, the pressure is suddenly released. This process of pressure and release widens the cracks and weakens the rock, causing it to eventually break apart. Some waves, especially those created by storms, are very high and forceful. In places where wave action is strong, the waves pick up the particles of rock and sand and throw them against the shoreline with a crashing motion. This produces a cutting action.

Some of the chunks and particles eroded from a shoreline may then be carried out to sea by waves. The particles sink and become deposited on the ocean floor. Other particles may be carried by longshore currents farther along the coast and deposited where there is shelter from the wind, and the wave action is not as severe. Even boulders may be carried by this means.

Erosion and deposition can change the geography of a shoreline over time. For example, 800 years ago, Hastings, in southern England, was on the coast. After centuries of deposition, it now lies several miles (kilometers) inland.

SHORELINE SURFACES

Seashore surfaces are classified as rocky, sandy, or muddy, depending upon their composition (make-up).

Rocky shores Rocky shores may consist of vertical cliffs, sloping shores, platforms, and boulder-covered areas. Vertical cliffs have no protection from the waves. However, on shores covered by boulders, spaces among the stones are protected, and pools of water, called tidepools, may form in them.

Beaches along rocky shorelines usually consist of pebbles and larger stones. The waves have carried away the finer particles.

Sandy shores Sandy shores make up about 75 percent of the world's seashores that are not ice-covered. They are constantly changing, depending upon the movement of the wind, the water, and the sand.

Some sandy shores form a steep slope down to the sea. On this type of shoreline, the waves break directly on the beach. Cycles of erosion and deposition are more extreme and have a greater effect on the shape of the shoreline. The greater the force of the waves, the steeper the slope, and the larger the sand particles.

Sandy shores that slope gently down to the sea are usually protected in some way from the full force of the water. A reef, for example, may have formed some distance from shore. As a result, the waves that reach the beach are gentle, erosion is slow, and sand particles are finer.

The texture of the sand helps determine what the beach will be like. Fine sand packs down and produces a smooth, gentle slope. Coarse sand allows the waves to sink in and move the particles around, producing a steep surface.

Muddy shores/estuaries An estuary is where a river, traveling through lowlands, meets the ocean in a semi-enclosed channel or bay. In these gently sloping areas, river sediments (soil and silt) collect, and muddy shores form. The water in an estuary is brackish—a mixture of fresh and saltwater.

Because estuaries are semi-enclosed by land, they are usually protected from the full force of ocean waves. High tide, however, may bring in a new supply of sea water and the water becomes more salty. During low tide, fresh water from the river dominates. After a rain washes soil into the river, the water carried into the estuary may be cloudy from the added sediments.

The sandy Ravenglass Estuary snakes past grassy fields on its journey. (Reproduced by Corbis. Photograph by Genevieve Leaper.)

LANDFORMS

Landforms found at the seashore include cliffs and rock formations; beaches and dunes; deltas; spits and bars; and certain types of saltwater wetlands.

Cliffs and rock formations High cliffs usually occur when highlands meet the sea. Pounding waves may gradually eat into the base of the cliffs, eroding the rock and creating a hollowed-out notch. Eventually, the overhang collapses and falls onto the beach, creating a platform of rock and soil.

Many coastlines consist of both hard and soft rock. Wave action erodes the soft rock first, sometimes sculpting strange and beautiful shapes, such as arches. Caves may gradually be carved into the sides of cliffs, or headlands may be created. A headland is a large arm of land made of hard rock that juts out from the beach into the ocean after softer rock has been cut away.

Beaches and dunes Beaches are almost-level stretches of land along the water's edge. They may be covered by sand or stones. Sand is just small particles of rock less than 0.079 inch (2 millimeters) in diameter. It may be white, golden, brown, or black, depending upon the color of the original rock. Yellow sand usually comes from quartz and black sand from volcanic rock. White or pink sand may have been formed from limestone, seashells, or coral particles.

Sand is carried not only by water but also by wind. When enough sand has been heaped up to create a ridge or hill, it is called a dune. Dunes range from 15 to 40 feet (4.6 to 12.2 meters) in height, and a few may grow as tall as 75 feet (23 meters). Individual dunes often travel, as the wind changes their position and shape. They tend to shift less if grasses take root in them and help hold them in place.

Deltas In estuaries, where rivers meet the sea, huge amounts of silt (very small particles of soil) carried from far inland by the river are deposited in the ocean along the shoreline. Large rivers can dump so much silt that islands of mud build up, eventually forming a fan-shaped area called a delta. In regions where the Earth's crust is thin, the weight of these sediments may cause the shoreline to sink. Huge deltas have formed at the mouths (where a river empties into a larger river or ocean) of the Mississippi River in the United States and the Nile River in Egypt.

Spits and bars A spit is a long, narrow point of deposited sand, mud, or gravel that extends into the ocean. In an estuary, a spit may reach up into the river's mouth.

BEWARE THE LUSCA!

Along the shores of the Bahama Islands, in the depths of the beautiful blue waters of the Caribbean Sea, lives the legendary Lusca, a creature that is supposedly half shark and half octopus. The Lusca draws people and even boats into its lair and, afterwards, like a rather rude dinner companion, it signals its satisfaction with a sudden upwelling "burp" of water.

Although not really a sea monster, the Lusca can be a bit tricky to get to know. A Lusca is not a creature at all, but an underwater cave carved out of the limestone that lies under the Bahama Islands. As sea levels rise and fall, these caves fill with water. When rainwater mixes with the seawater, their different densities (weights) cause whirlpools. These whirlpools account for people being drawn down into the caves and for the sudden burps of water. These extreme conditions often churn up sediments, making the caves dangerous for divers who, in the resulting dimness, may not be able to find their way out again.

A bar is an underwater ridge of sand or gravel formed by tides or currents that extends across the mouth of a bay (an area of the ocean partly enclosed by land). If the bar closes off the bay completely, the water trapped behind the bar is called a lagoon. A tombolo is a bar that has formed between the beach and an island, linking them together.

Saltwater wetlands Saltwater wetlands are portions of land covered or soaked by ocean water often enough and long enough to support plants adapted for life under those conditions.

A swamp is a type of wetland dominated by trees. A saltwater swamp is formed by the movement of the ocean tides. When the tide is low, flat places in the swamp are not under water.

Marshes are wetlands dominated by nonwoody plants such as grasses, reeds, rushes, sedges, and rice. Saltwater marshes are found in low, flat, poorly drained coastal areas. They are especially common in deltas, along low-lying seacoasts, and in estuaries. Saltwater marshes are greatly affected by the tides, which raise or lower the water level on a daily basis. A saltwater marsh may have tidal creeks, tidal pools, and mud flats.

For a more complete discussion of saltwater swamps and marshes, see the chapter titled "Wetland."

Sand dunes along the coast of Lake Michigan. Once grasses take root in dunes, they are less likely to travel with the wind and change their position and shape. (Reproduced by permission of National Park Service. Photograph by Richard Frear.)

PLANT LIFE

Seashore plants include microscopic, one-celled organisms, many kinds of seaweeds and seagrasses, and even some species of trees. The types of plants that grow in a particular region are determined by climate and the kind of shoreline surface—rocky, sandy, or muddy. Some sandy or muddy shores do not support plants because the soil is frequently disturbed.

Plants that live in the lower (sublittoral) zone along the seashore are surrounded by water at all times. For this reason, most have not developed the special tissues and organs for conserving water that are needed by plants on land. The surrounding water also offers support to these plants, helping to hold them upright. As a result, their stems are soft and flexible, allowing them to move with the currents without breaking.

Plants that live in the intertidal (littoral) zone must be the hardiest, because they are exposed to the air for part of the day and are underwater for part of the day. Plants in the upper (supralittoral), or splash, zone have adapted to life on land. However, they must be tolerant of salty conditions.

Plants that live along the seashore can be divided into four main groups: algae (AL-jee), fungi (FUHN-jee), and lichens (LY-kens), and green plants.

ALGAE, FUNGI, AND LICHENS

Most marine plants are algae. (Although it is generally recognized that algae do not fit neatly into the plant category, in this chapter they will be discussed as if they were plants.) Most algae have the ability to make their own food by means of photosynthesis (foh-toh-SIHN-thih-sihs; the process by which plants use the energy from sunlight to change water and carbon dioxide from the air into the sugars and starches they use for food). Algae also require other nutrients that they obtain from the water. In certain regions, upwelling of deep ocean waters during different seasons brings more of these nutrients to the surface. This results in sudden increases in the numbers of algae. Increases also occur when nutrients are added to a body of water by sewage, or by runoffs from fertilized farmland.

Low tide on the Strait of Juan de Fuca exposing seaweeds. (Reproduced by permission of Corbis. Photograph by David Muench.)

Some forms of algae are so tiny they cannot be seen without the help of a microscope. Called phytoplankton (fy-toh-PLANK-tuhn), they float freely in the water, allowing it to carry them from place to place. Other species, often referred to as seaweeds, may be anchored to the seafloor.

Fungi cannot make their own food by means of photosynthesis. Some, like mold, obtain nutrients from dead or decaying organic matter. They assist in the decomposition (breaking down) of this matter and in releasing many nutrients needed by other plants. Others are parasites and attach themselves to, and feed on, other living things.

Lichens are combinations of algae and fungi that live in cooperation: the fungi surround the algal cells, and the algae obtain food for themselves and the fungi by means of photosynthesis. Lichens prefer the upper zones of the seashore, surviving dryness by soaking up water during high tide.

Growing season Algae contain chlorophyll, a green pigment used to turn energy from the Sun into food. As long as light is available, algae can grow. In some species, the green color of the chlorophyll is masked by orange-colored compounds, giving the algae a red or brown color.

The growth of ocean plants is often seasonal. In northern regions, the most growth occurs during the summer. In temperate (moderate) zones, growth peaks in the spring but continues throughout the summer. In regions near the equator, growth is steady throughout the year.

Reproduction Algae may reproduce in one of three ways. Some split into two or more parts, each part becoming a new, separate plant. Others form spores (single cells that have the ability to grow into a new organism). A few reproduce sexually, during which cells from two different plants unite to bring forth a new plant.

Fungi and lichens usually reproduce by means of spores. Lichens can also reproduce when soredia (algal cells surrounded by a few strands of fungus) break off and form new lichens wherever they land.

Common algae Two types of algae commonly found along the seashore include phytoplankton and seaweeds. Diatoms and dinoflagellates (dee-noh-FLAJ-uh-lates), are the most common types of phytoplankton.

Seaweeds are forms of green, brown, and red algae that grow primarily along rocky seashores. Green algae prefer the upper zone where they are exposed to sunlight and fresh water from rain. Brown algae prefer the middle zone and shallow water. They absorb sunlight readily and they are tough enough to endure the action of waves and tides. Red algae live in tidepools or offshore waters, as do the large forms of seaweed, called kelps.

Seaweeds are different from land plants in that they do not produce flowers, seeds, or fruits. They also lack root systems, for they do not need roots to draw water from the soil. They do, however, have rootlike holdfasts which anchor them to rocks. They are seldom found on sandy shorelines because there are no rocks for anchorage.

The type of shoreline determines the species of seaweeds that will grow there. Also, different species prefer different zones along the same shoreline. Some are better adapted to dry conditions than others. Some species, for example, have a thick layer of slime that prevents water loss, while others have a layer of tissue that retains water.

GREEN PLANTS

Green plants have roots, stems, leaves, and often flowers. Most green plants need several basic things to grow: light, air, water, warmth, and nutrients. Like algae, green plants depend upon photosynthesis for food. The remaining nutrients—primarily nitrogen, phosphorus, and potassium—are obtained from the soil. These minerals may not be in large supply. In some seashores along dry, desert coasts, plants have evolved that can absorb mist from the nearby ocean through their leaves. The mist provides enough water for them to survive.

Green plants grow in all seashore zones. Those that grow in the lower zone are sea plants, and those that grow in the upper zone are land plants.

Growing season The growing season depends upon where the seashore is located. Near the equator, growth continues throughout the year. In northern regions, there is a spurt of growth during the summer. In moderate climates, growth begins in the spring and continues throughout the summer.

Reproduction Green plants often depend upon the wind and insects to carry pollen from the male part of a flower, called a stamen, to the female part of a flower, called a pistil, for reproduction. This process is called pollination. Others send out stems from which new plants sprout.

Common green plants Green plants found along the seashore include seagrasses, such as eelgrass and paddleweed. These plants live in the lower zone but are similar to land grasses. They have roots, and they bloom underwater. Beds of seagrass occur in sandy or muddy shores in calm areas protected from currents, such as in lagoons or behind reefs. They attract a wide variety of grazing marine animals. These grasses help slow the movement of the water and prevent erosion of the shoreline.

The high salt content of the seawater in the intertidal zone makes it hard for land plants to adapt. However, grasses such as marsh grass, cord grass, salt hay grass, and needlerush can thrive here. In salt marshes, glasswort and sea lavender are also found.

While sand dunes are too dry to support much plant life, there are many species of plants that grow on dunes. These include sandwort, beach pea, marram grass, yellow horn poppy, beach morning glory, and sea oats.

Sea oats grow a long taproot (main root) that may extend more than 6 feet (1.8 meters) into the sand to reach water. Dunes may also support trees, such as pine and fir, if the soil is stable and fresh water is available.

Mangrove trees grow in estuaries in tropical zones where the shoreline is muddy. Some are small, shrublike plants; others are tall and produce large forests. Mangroves have two root systems. One is used to anchor them in the muddy bottom. The other is exposed to the air from which it pulls oxygen, for the soil in mangrove swamps is usually low in oxygen. (For more information about saltwater wetland plants, see the chapter titled "Wetland.")

ENDANGERED SPECIES

Because coastlines are popular places for people to live, plants can suffer when their natural habitat is disturbed. Algae and seagrasses can be destroyed by polluted water. Dune grasses are easily trampled by beachgoers or destroyed by dune buggies and other off-road vehicles. Also, sometimes visitors pick flowering plants, which limits the plants' ability to reproduce.

ANIMAL LIFE

The kinds of animals that live along a particular shoreline are determined by zone—upper, intertidal, or lower—and the type of surface—rocky, sandy, or muddy. This is also true of different species of the same animal. Hermit crabs, for example, live comfortably in the lower and intertidal zones of rocky shores, whereas mudcrabs prefer muddy estuaries.

Lower zone animals, including sea anemones, shrimp, and small fish, are underwater almost all of the time. Upper zone animals, such as ghost crabs, prefer dry conditions and therefore live on land. The middle, or intertidal, zone along rocky shores supports the most lifeforms, in spite of its being the harshest zone of all. Cockles, barnacles, clams, sea urchins, and fish can all be found in this zone. Some, such as clams, can survive during low tide when they are exposed to air by closing their shells to keep from drying out. Their shells also help them survive battering by the waves. Others find shelter in small pools where water collects in depressions and among rocks when the tide goes out, remaining there until high tide returns.

On depositional (sandy or muddy) shores, animals are constantly shifted around by the motion of water because there is nothing on which to anchor themselves. Many burrowing animals, such as clams and lugworms, survive by digging into the sand. They often have siphons (tubes) through which they can draw in oxygenated water and food.

Life on eroding shores is even harder; animals must attach themselves firmly to rocks or be washed away by the battering waves. The crevices of the rocks are homes to soft-bodied animals such as sea anemones. Dark caves found on many rocky shorelines provide a place in which many sea creatures may take shelter.

MICROORGANISMS

Seashores are home to many kinds of microscopic animals. Most microorganisms are zooplankton (tiny animals that drift with the current) that live on the surface of the sea. They include protozoa, nematodes (worms), and the larvae or hatchlings of animals that will grow much larger in their adult form. Some zooplankton eat phytoplankton and, in turn, are preyed upon by other carnivorous (meat-eating) zooplankton.

Bacteria Bacteria are found in every zone along the shore and provide food for lower animals forms. They also help decay the dead bodies of larger organisms.

Because much of the water that washes up on a beach sinks downward again, the sand or gravel acts as a filter. Particles of matter suspended in the water become trapped between the grains of sand. These particles become food for the bacteria which consume them and then release other nutrients into the water.

INVERTEBRATES

Animals without backbones are called invertebrates. Many species are found along the seashore. Crustaceans and mollusks, invertebrates that usually have hard outer shells, are well adapted to the intertidal zone because their shells help prevent them from drying out during low tide. Soft-bodied animals, such as sea anemones and small octopi, prefer rocky shores where they can hide in rock crevices and survive low tide in rock pools.

Sand dunes are home to many insects, such as wasps, ants, grasshoppers, and beetles. Some insects and spiders burrow into the sand during the heat of the day and hunt for food at night.

Food Invertebrates may eat phytoplankton, zooplankton, or both. Mollusks draw sea water in through their siphons and filter out the tiny creatures, which they then consume. Some invertebrates also eat plants or larger animals. Starfish, for example, dine on mollusks.

> ### THE WEB OF LIFE: TAKE A NUMBER AND GET IN LINE!
>
> Life exists everywhere on our planet, even in the most remote, unfriendly places. Often, one life form leads the way for others because its presence in the habitat causes changes. Soon, other life forms move in, some to eat the first inhabitants, others because the habitat is now more comfortable for them. This process of succession occurs regularly along the seashore in some surprising ways.
>
> Suppose you have bought a new boat. You launch it and tie it to the end of a pier. Before your back is turned, life forms will have begun to build colonies on its hull. Within an hour, bacteria will attach themselves to any surface below the water line. Phytoplankton and zooplankton come next, usually within the first day. By the second or third day, hydroids and bryozoa, tiny but more complex animals, move in too. If the boat is not moved, barnacles and algae will have attached themselves to its hull by the end of the week and the other animals will have been greatly reduced in number. Eventually, mussels will move in, crowding everybody else out.

Reproduction Most invertebrates reproduce by means of eggs. In some cases, a parent watches over the young in the early stages. In other instances, the young are not cared for by the parents. As a result, survival often depends on the absence of predators.

Common seashore invertebrates Tiny crustaceans called sand hoppers, or beach fleas, live on sandy beaches where they can hop as high as 1 foot (30 centimeters) into the air. They are not real fleas and do not bite humans or other animals but live on seaweed and dead matter.

On rocky shores, many species of periwinkles are common. Periwinkles are snails that come in many colors, including brown, yellow, and blue. They eat algae and can live many days without water. A periwinkle has a hinged door in its shell that it can shut, keeping moisture locked inside.

On muddy shores, mangrove forests are often homes for oysters and barnacles, which attach themselves to the trunks and roots of trees. Snails and crabs are common in saltwater marshes.

REPTILES

Reptiles are cold-blooded vertebrates such as lizards and snakes. Only one species of lizard, the marine iguana, lives both in the sea and on the seashore. However, pine lizards may be found in dry areas on sandy beaches

Starfish and anemones in Cape Kiwanda tidepool. Both animals are common invertebrates to the seashore. (Reproduced by permission of Corbis. Photograph by Stuart Westmorland.)

where pitch pines grow. Some species of snakes, such as the hognose snake, live among sand dunes on sandy shorelines, preferring the dry conditions found there. Fowler's toads can also be found on upper beaches. A saltwater crocodile makes its home in the waters of Southeast Asia, where it lives along muddy seashores near the mouths of rivers. But the reptile most commonly found at the seashore is the sea turtle.

Food Turtles eat soft plants, as well as small invertebrates such as snails and worms. Turtles have no teeth. Instead, they use the sharp, horny edges of their jaws to shred the food enough so they can swallow it.

Sand dune snakes are carnivorous and catch mice and other small prey. They also eat the eggs of nesting shorebirds. Saltwater crocodiles are also carnivorous and feed primarily on fish and turtles. They have been known to attack and eat humans.

Reproduction Both snakes and turtles lay eggs. Turtles lay theirs in a hole on a sandy shore, which they then cover over with sand. After the nest is finished, the female abandons it, taking no interest in her offspring. Six weeks later the eggs hatch and the young turtles make a run for the ocean and disappear.

Common seashore reptiles Sea turtles can be distinguished from land turtles because they have flippers, instead of toes, which enable them to swim, and by their tolerance for salt water. Green sea turtles are migrators, traveling as far as 1,250 miles (2,000 kilometers) to return to a particular breeding area where they lay their eggs. (More on green sea turtles can be found in the chapter titled "Ocean.")

The saltwater, or estuarine, crocodile lives in northern Australia and in the region stretching from the east coast of India to the Philippines. It is one of the largest species of crocodiles, and average males grow between 12 and 15 feet (3.6 to 4.5 meters) long. Most crocodiles float on the surface of the water close to the shoreline where they wait for prey to wander past. The saltwater crocodile is the only species known to swim out to sea.

> ## BURSTING AT THE SEAMS
>
> What happens when a crab needs a new suit of armor? Crabs, like other crustaceans, have a hard outer shell. But this shell does not grow as the crab grows. Any crab that outgrows its current shell has to get rid of it fast or feel the pinch. This shedding of the shell is called molting.
>
> Molting is caused by hormones, chemical messengers in the crab's body that help split the shell so the crab can climb out. While it runs around shell-less, the crab is in greater danger from predators. But eventually its skin hardens and soon it has a new, ready-made suit of armor.

FISH

Fish are primarily cold-blooded (depending on the environment for warmth) vertebrates having gills and fins. The gills are used to draw in water from which oxygen is extracted. The fins are used to help propel the fish through the water.

Most fish found along the seashore are smaller varieties, such as rock bass, that can live in shallow water or in tidepools. They are dull in color so that they match the sand or gravel background. Some have suckers (suction cups) on their undersides that allow them to cling to rocks so they will not be washed away. The exception to these rules is the flounder, which can grow to be quite large and which spends its entire life in the shallow pools of water on the beach. Fish are discussed in greater detail in the chapter titled "Ocean."

Food Some fish, such as the clingfish of Chile, attach themselves with suckers to rocks and scrape off any animals or plants on which they feed before letting the waves carry them back out to sea. Fish that swim in search of prey, such as some sculpin species, may lie in wait in rockpools until the tide brings them something tasty. Most shoreline fish are slow swimmers that feed on other slow-moving creatures, such as shrimp.

Reproduction Fish reproduce by laying eggs. Some build nests and care for the new offspring, while others carry the eggs with them until they hatch, usually in a special body cavity or in their mouths. For example, female seahorses lay their eggs in a pouch on the underside of the male, who then carries the eggs until they hatch.

Some fish, such as the Atlantic menhaden, live in deep waters for most of the year. Then, during breeding season, they appear in large numbers in shallower shoreline waters where they lay their eggs.

Common seashore fish Flatfish live in shallow water on sandy shorelines. They begin life looking like other fish; however, as they mature, they lie on one side and become flattened. The eye on the "bottom" side migrates to the top until both eyes are looking upward.

Mudskippers are small amphibious (am-FIB-ee-yuhs) fish that live along muddy shores, especially where mangrove forests flourish. Amphibious means that they can live either on land or in water. When the tide is in, mudskippers swim and breathe underwater. When the tide is out, they breathe air and use their fins to hop among the tree roots.

Other shoreline fish include spiny dogfish, which are a species of shark, common anchovies, killifish, northern pipefish, bluefish, northern

---

PLASTIC IS NOT A NUTRIENT!

Sea turtles love to snack on jellyfish, and what resembles a jellyfish when puffed up with air? A plastic bag. Unfortunately, a sea turtle who can't distinguish between the two, can die. Because plastic bags can't be digested and moved through the animal's body, the bags and other plastic garbage block an animal's digestive tract. Dolphins, seals, whales, and other animals are also in danger from plastics. A gallon-sized plastic milk bottle, a plastic float, a garbage bag, and other items were found wedged in the digestive tract of a dead sperm whale.

How does plastic become a problem? People litter. In 1993, more than 224,000 people in 55 countries volunteered to clean up more than 5,000 miles (8,000 kilometers) of beach. They collected a total of 5,000,000 pounds (2,300,000 kilograms) of trash, much of it plastic, including bags, caps, Christmas trees, and even Barbie dolls.

barracuda, striped bass, cowfish, four-eyed butterfly fish, northern sea robins, barred sea perch, and kelp greenlings.

BIRDS

Birds are vertebrates. Most seabirds remain near land where they can nest during breeding season. Many have adapted to marine environments by means of webbed feet for swimming and special glands for removing excess salt from their blood.

Shorebirds, such as oystercatchers and gulls, feed or nest along the coast and prefer hunting in shallow water. Their feet are usually webbed and most migrate. Lesser golden plovers, for example, spend their winters in South America and their summers in the Arctic. During their travels, they may stop to rest on the eastern coast of the United States.

Wading birds, such as the American avocet, are often found along muddy shores. They have long legs, wide feet, long necks, and long bills that are useful for nabbing fish and other food.

A waterfowl is a bird that spends most of its time on water, such as a duck. Their legs are positioned closer to the rear of their bodies, which is good for swimming, but awkward for walking. As a result, they waddle as they walk. Their bills are designed for grabbing vegetation, such as grasses, on which they feed.

A large number of newly hatched baby sea turtles. Soon they will make a run for the ocean and disappear. (Reproduced by permission of Corbis. Photograph by Robert Pickett.)

Farther inland, common land birds, such as red-winged blackbirds, may live among dune grasses where they hunt for insects.

Food All birds that live along the seashore are carnivorous. Most eat fish, squid, or zooplankton, and they live where food is plentiful. Several species, such as eiders, dive to the bottom of shallow water where they feed on shrimp, worms, or crabs. Other species, such as sanderlings, dash into the water as waves retreat in order to grab prey before it has a chance to get away. Some, like gulls, are scavengers that eat dead matter that other creatures have discarded.

Reproduction Sea and shorebirds usually nest on land. Some nest in huge colonies on the ground, others dig burrows, and still others, like the kittiwakes, prefer ledges on a cliff. Like ordinary land birds, they lay eggs and remain with the nest until the young are able to survive on their own. Some live and feed in one area and migrate to another for breeding. Birds that nest on sandy shores tend to lay speckled or blotchy eggs in beige and brown colors that blend in with the sand and pebbles in order to protect the eggs from predators.

Flatfish, like this one, are common to the seashore. (Reproduced by permission of Corbis. Photograph by Brandon C. Cole.)

Common seashore birds Birds that live along the seashore include gulls, plovers, pelicans, boobies, puffins, ruddy turnstones, sandpipers, sanderlings, and penguins. There are only a few true sea ducks, including the eider and the scoter.

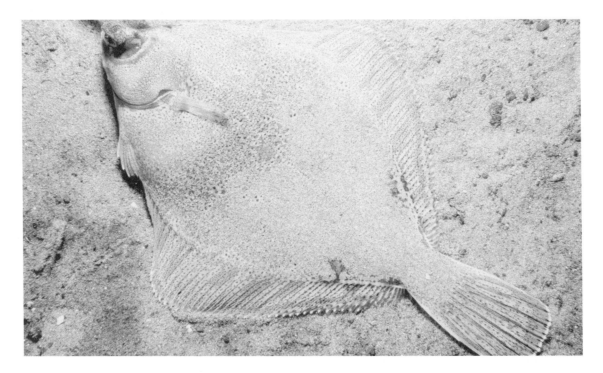

MAMMALS

Mammals are warm-blooded vertebrates that have at least some hair and bear live young. Few mammals live permanently along the seashore. Of those that do, most, such as mice, are small and prefer dry areas out of reach of the tide. However, many land animals, such as skunks, rats, foxes and raccoons, may visit the shoreline at night in search of food. They prefer low tide, when mussels, oysters, worms, and tidepool animals are exposed and easy to find.

Food Many mammals are carnivorous. Raccoons, for example, hunt crabs, shrimp, and turtle eggs, while sea otters often dine on mussels and other invertebrates, using rocks and other "tools" to break the shells. Some mammals, such as mice, are omnivorous, which means they eat both plants and animals, such as insects. Others, like rabbits, eat only plants.

Reproduction Mammals bear live young. While most have only one offspring at a time, rodents, such as mice, have large families. The young are nursed with milk produced by the mother's body until they are able to find food on their own. This is true whether the mammal spends some of its time in the water or all of the time on shore.

Common seashore mammals Sandy beaches are home to cottontail rabbits, voles, and mice. Rocky shores in some regions provide breeding and socializing areas for seals and walruses. Sea otters are also found along rocky coastlines. Along muddy shores that have marshy areas, muskrats and manatees are common.

ENDANGERED SPECIES

Certain species of sea turtles are endangered. Because sea turtles lay their eggs on beaches, the eggs are easily hunted and/or destroyed. The eggs are a popular food in many parts of the world, as are the turtles themselves. Turtles are also threatened by pollution, which can destroy their food supply.

In certain areas, other animals are also endangered. They include the starlet sea anemone, the osprey, the Eskimo curlew, the dalmatian pelican, and the West Indian manatee.

HUMAN LIFE

In times past, the oceans were the only transportation route between the eastern and western hemispheres. Also, the oceans have always been important for fish and other food resources. For these reasons, people have tended to live within an easy distance of the seashore. As of 1994, more than 60 percent of the world's population—about 2.5 billion people—lived in or around cities located on the coast. That number is expected to jump to 6.3 billion by the year 2028.

IMPACT OF THE SEASHORE ON HUMAN LIFE

Because people have always lived along the seashore, it has had an important effect on human life.

Food Various species of seaweed are used by people as food and in food preparations. Humans also eat many of the animals that are found along shorelines, including crabs, mussels, clams, oysters, shrimp, fish, turtles, and turtle eggs. Anchovies, tiny fish that live in shallow coastal waters, are one of the most important commercial fishes.

Fish farming, the raising of fish for commercial purposes, is also an important industry of the coastlines. For more information about fish farming, see the chapter titled "Continental Margin."

Harbors Many of the world's great cities—New York, San Francisco, Hong Kong, Tokyo, London, to name a few—are harbor cities. From these harbors, millions of tons of goods are shipped around the world, which has an important effect on the world economy. Therefore, many factories are built near

Monaco City and its harbor on the Mediterranean coast. Millions of tons of goods are shipped from harbors such as this. (Reproduced by permission of Corbis. Photograph by Jonathan Blair.)

harbors in order to keep transportation costs low. Passenger ships and ferries also use these harbors to carry people from place to place.

Energy The ocean is a source of energy. The energy in the tides, for example, can be harnessed to produce electrical power. The first tidal power station was developed in an estuary in France in 1966. Turbines (engines with fanlike blades) are built into a dam that spans the estuary. As the tides flow in and out of the estuary, they turn the turbine blades. Plans have been made to build a similar station on the estuary of the Severn River in England.

Other types of power stations may also be located by the sea, because seawater is used to cool their machinery.

Minerals and metals Minerals and metals are other important resources of coastlines. Rocks, sand, and gravel dredged from the sea bottom, especially off the coasts of England and Japan, are used in the construction of roads and buildings. Along the Namibian coast of southwest Africa, diamonds are mined.

Recreation For many people, nothing is as much fun as a day at the beach, and sandy beaches in particular attract millions of visitors each year. Huge hotels, resorts, and amusement parks have been built along coastlines where people can enjoy sailing, swimming, fishing, walking, camping, and just lying in the sun.

SURF'S UP!

Next to swimming and sunbathing, surfing is probably the best-known beach activity. Surfing involves being carried on the sloping portion of a wave as it moves toward shore. It is especially popular along the coasts of Hawaii, California, New Zealand, Australia, South Africa, Puerto Rico, Peru, Great Britain, and Brazil, where wave conditions are the most favorable.

In "body" surfing, the person first swims toward shore, matching the speed of a particular wave. The swimmer then stiffens the body in order to provide a flat surface that will glide on the front of the wave and carry him or her to shore. To surf using a board, the person lies on the board and paddles until the board is going as fast as the chosen wave. The person then stands up and "rides" the wave toward shore.

Surfboards weigh between 24 and 40 pounds (11 and 18 kilograms) and are 6 to 12 feet (1.8 to 3.7 meters) long. Surfing requires good balance, timing, and coordination because boards can reach speeds of 35 miles (56 kilometers) per hour.

Surfing originated in Hawaii, where it was used by the nobility as part of a religious ceremony. In 1920, Duke Kahanamoku of Hawaii, the Olympic swimming champion, formed the first surfing club and almost singlehandedly made the sport popular.

IMPACT OF HUMAN LIFE ON THE SEASHORE

While the benefits that coastlines have brought to human life are many, human life has brought many impacts to the seashores.

Use of plants and animals After World War II (1939–45), the technology of commercial fishing improved and a growing population increased the demand for fish as a food source. By the 1970s, major food species, including anchovies and certain shellfish, had been greatly reduced.

Fish farms have helped maintain certain species of commercially popular fish. The fish most commonly farmed are shellfish, such as oysters, mussels, scallops, and clams. Crabs, lobsters, shrimp, salmon, trout, and tilapia are also farm raised, but to a lesser extent. The output of fish farms has tripled since 1984, and it is estimated that, by the year 2000, the numbers of fish produced by fish farms will, in some cases, exceed those caught in the wild.

Other sea plants and animals have been taken as souvenirs or art objects, reducing their numbers. When sea shells are taken from dead animals, no harm is usually done. However, many shells available commercially are taken from living animals and the animals are then left to die.

Preventing natural changes Homes and other structures built along coastlines are threatened when storms, erosion, and other natural processes change the seashore. Sometimes, people try to work against nature by building walls and other barriers to protect these structures. Often, however, these efforts make the problems worse. As the natural shape of a shoreline is changed, waves may become stronger and do even more damage.

Changes made by humans to protect one shoreline may also cause unexpected, undesirable changes in another. Walls built to prevent the sand from being washed away from beaches at Ocean City, Maryland, for example, kept that sand from naturally rebuilding the beaches on Assateague Island to the south. Since the walls were built, damage from erosion on Assateague has increased from 2 feet (.6 meter) to 36 feet (11 meters) each year.

Overdevelopment Use of the land for harbors, recreation, and housing has changed the appearance of many shorelines. Because beaches are so popular, many have become crowded with so many people and buildings that wildlife has either been destroyed or frightened away. In many places, the coastline is no longer a place to go to "get away from it all." It has become just another part of a big city.

Dune buggies, dirt bikes, and other recreational vehicles destroy plants and scare animals. Hikers can also destroy vegetation, and even well-meaning people who wish only to observe nature can frighten animals or upset the natural rhythms of their lives.

Quality of the environment In 1994, there were 2,279 occasions when beaches were closed in the United States because the water was polluted. Most ocean pollution caused by humans is concentrated along the seashore. Sewage and industrial wastes are dumped from coastal cities, adding metals and chemicals to the water. Discarded items, such as plastic bags and food wrappers, also pose health hazards for both animals and people.

Insecticides (insect poisons) and herbicides (weed poisons) reach the oceans when the rain washes them from fields and they are carried by rivers to the sea. These poisons often enter the food chain and become concentrated in the bodies of some organisms. Fertilizers and human sewage are also a problem. They cause phytoplankton to reproduce rapidly. When the plants die, their decaying bodies feed bacteria. The bacteria reproduce and use up the oxygen in the water, and other organisms, such as fish, soon die.

Industrial accidents and waste are other problems. Between 1953 and 1960, more than 100 people died and 2,000 were paralyzed when they ate fish contaminated with mercury, a metallic element used at a nearby factory at Minamata Bay in Kyushu, Japan. Oil spills from tanker ships are another danger, as is oil from oil refineries, pipelines, and offshore oil wells. Power plants and some industries also often dump warm water into the oceans, causing thermal (heat) pollution. Organisms that require cooler water are killed by the increase in the water temperature.

Agriculture, construction, and the removal of trees digs up the soil. The rain then washes the soil into streams and rivers. Eventually, it enters the ocean to collect as sediment in coastal areas. Some organisms that cannot survive in heavy sediment, such as clams, then die.

Pollution can travel, so that problems caused by one country may damage the shoreline environment in another. For example, it is estimated that 30 percent of all the litter on Scotland's beaches comes from the east coast of North America.

> ## THE WORLD'S WORST OIL SPILL
>
> Oil spills are especially hazardous to coastlines. The oil is carried ashore by ocean waves and tides, but while the water retreats back to the sea, the oil remains. Thick deposits stick to the shoreline and coat the plants and animals that live there. The oil mats the feathers of birds and the fur of swimming animals so that the creatures cannot keep warm, and many die of the cold.
>
> The world's worst oil spill took place in 1989 when the ship *Exxon Valdez* was wrecked in Prince William Sound, Alaska. Eleven million gallons (42 million liters) of oil were spilled. Over time, the oil spread out, covering 10,000 square miles (25,900 square kilometers) of ocean and 1,200 miles (1,940 kilometers) of coastline. It is estimated that as many as 6,000 otters, whales, dolphins, and seals, and 645,000 birds died. Three nearby national parks and three national wildlife refuges were also affected.

NATIVE PEOPLES

In most parts of the world, early peoples often settled near coastlines where fish were plentiful and the ocean offered a means of transportation. Since the mid-twentieth century, desirable seashore locations have been taken over by tourists and non-native residents. Also, few native peoples have continued to live traditional lifestyles. Many have gone to live in cities or adopted more contemporary ways of life. Among those that have tried to maintain important elements of their traditional cultures are the Samoans and Native American tribes of the northwest coast.

Samoans The Samoa Islands lie in the Pacific Ocean southwest of Hawaii. The largest island, Savai'i, has an area of only 602 square miles (1,565 square kilometers), so the seashore is easily accessible and an important part of everyone's life.

The people of Samoa are Polynesian and are closely related to the native peoples of Hawaii and New Zealand. Although the islands have felt the influence of Europe and then America since 1722, traditional Samoan culture has remained very important.

The Samoan economy is based on agriculture, and crops include corn, beans, watermelons, bananas, and pineapples. Although fish have always been an important food source for local families, fishing was not commercially important until 1953 when a tuna cannery was built in the town of Pago Pago.

Native Americans of the northwest coast The Columbia and Fraser Rivers in North America open into the Pacific Ocean, and each year millions of Pacific salmon return to those rivers to breed. This ample supply of fish caused many native peoples to settle there, including the Tlingit, Haida, Tsimshian, Kwakiutl, Nootka, Salish, Hupa, Yurok, and Karok. The area

---

KON-TIKI

Norwegian explorer Thor Heyerdahl (1914–) was convinced that ancient peoples once sailed across 4,300 miles (6,900 kilometers) of the Pacific Ocean from Peru and colonized Polynesia. To prove his theory, in 1947 he set out with a crew of six people to duplicate the feat. The expedition, which lasted 101 days, was carried out on a large raft called the *Kon-Tiki*.

Named for a legendary Inca Sun-god, the *Kon-Tiki* was patterned after sailing rafts used by the ancient Incas, native peoples who lived in what is now Peru. The raft measured 45 feet (13.7 meters) long in the center and tapered to 30 feet (9.1 meters) at the sides. No metal was used. Instead, the nine thick balsa logs were tied together with hemp rope. Two masts supported a large rectangular sail, and a bamboo cabin was built in the center for shelter.

On April 28, 1947, the raft was set adrift 50 miles (80 kilometers) off the coast of Peru. Ninety-three days later, the *Kon-Tiki* sailed past the island of Puka Puka, which lies east of Tahiti. On August 7, 1947, after the *Kon-Tiki* crashed into a reef in the Tuamoto Archipelago, the voyage was finished, but Heyerdahl had proved his point. The *Kon-Tiki* is now in a museum in Oslo, Norway. Heyerdahl wrote about the voyage in his book, *Kon-Tiki*.

---

proved rich in other wildlife as well. Mussels, clams, oysters, candlefish, herring, halibut, and sea lions were available from the sea. The land, too, provided food such as caribou, moose, mountain sheep and goats, deer, and many small animals, as well as roots and berries.

Products of the seashore, such as shells, dried fish, fish oil, and the dugout canoes built for transportation were considered a source of wealth. Religious beliefs were based on mythical ancestors whose images were carved on totem poles, boats, masks, and houses.

By 1900, traditional ways of life were disappearing, but the people remained on or near their ancient lands. Many now work in forestry. Some groups, however, are restoring native customs, and there is also interest in a return to arts-and-crafts production.

THE FOOD WEB

The transfer of energy from organism to organism forms a series called a food chain. All the possible feeding relationships that exist in a biome make

Western Samoan villagers sing and wave to visitors from their sailing canoe. (Reproduced by permission of Corbis. Photograph by Hulton-Deutsch Colle.)

SEASHORE PARKS AND RESERVES OF THE WORLD (PARTIAL LIST)

Name	Location	Square miles (square kilometers)	Features
Aldabra Islands Nature Reserve	Seychelles (Africa)	60 (156)	Protects giant tortoise and other animals
Bako National Park	Sarawak	10 (26)	Bays and coves; forest; wild pigs, deer, monkeys, birds
Cape Le Grand National Park	Australia	62 (161.2)	Beach; coastal plants
Easter Island National Park	Chile	66 (171.6)	Protects vegetation and animals
Elat Gulf Coral Reef Reserve	Israel	0.5 (1.3)	Coral reefs with tropical fish
Eldey Nature Reserve	Iceland	0.006 (0.015)	Gannet breeding area
Franklin D. Roosevelt National Park	Uruguay	58 (150.8)	Sand dunes, pines
Ise-Shima	Japan	201 (522.6)	Forested coastline; pearl farms
Kong Karls Land Reserve	Norway	200 (520)	Arctic islands; polar bears
Kranji Reserve	Singapore	0.08 (0.21)	Mangrove marsh
Kyzylagach	Azerbaijan	363 (943.8)	Coastal reed and salt marshes; flamingos, bustards
Pembrokeshire Coast	Great Britain	225 (585)	Rocky coast; island birds
Prince Edward Island	Canada (Gulf of St. Lawrence)	7 (18.2)	Forested coastline; many small mammals
Rikuchu Kaigan	Japan	45 (117)	Cliffs, islands, and beaches; forests; birds
Skallingen	Denmark	12 (31.2)	Dunes, marshes
St. Lucia	Natal (Africa)	190 (494)	Estuary on False Bay; hippopotami and birds
Westhoek Nature Reserve	Belgium	1.3 (3.4)	Sand dunes, marine plants
Yala National Park	Sri Lanka	91 (236.6)	Lagoons, rocky hills

Lighthouses are frequently found on seashores. They are used to guide boats into ports and docks. [Reproduced by permission of Corbis Corporation (Bellevue).]

A hermit crab is just one species of crustacean found along the seashore. They search the shallow water of the shore for food. [©1995 PhotoDisc].

A group of walruses gathered along a seashore. They use the shore to mate and take care of their young. [©1995 PhotoDisc.]

A deserted beach. Many of the world's beaches can no longer be used by humans or animals because of pollution. [©1995 PhotoDisc.]

Shells found along the shore are the abandoned shelters of animals such as snails. When the snails grow too large for one shell, they leave it to find a more spacious one. [©1995 PhotoDisc.]

The Monkton tidal bore in the Bay of Fundy in Canada. [Reproduced by permission of Corbis Corporation (Bellevue).]

Dog sleds are a common form of transportation for people living a traditional lifestyle on the tundra of Alaska. [Reproduced by permission of Planet Earth Pictures Limited.]

A caribou herd crossing a river on the Alaskan tundra. Caribou spend only the summer months on the tundra since the winter is too harsh. [Reproduced by permission of Planet Earth Pictures Limited.]

Mountain goats are hunted by alpine tundra dwellers for food. [©1995 PhotoDisc.]

Snow leopards are just one species of mammal commonly found in the tundra. [©1995 PhotoDisc.]

Alpine tundra dwellers often raise llamas for their milk and as a source of food. [Reproduced by permission of Planet Earth Pictures Limited.]

Icebergs and floating ice pieces, like these in Paradise Bay in Antarctica, are common sites in the tundra. [©1995 PhotoDisc.]

A cypress swamp. There are three main types of wetlands: swamps, marshes, and peatlands. [©1995 PhotoDisc.]

The carnivorous (meat-eating) pitcher plant commonly found in peatlands. [©1995 PhotoDisc.]

The American beaver, with its webbed feet, is well adapted to life in wetlands. [©1995 PhotoDisc.]

Reeds are plants that play a key role in wetland succession. [©1995 PhotoDisc.]

Alligators are fierce predators common to swamps and other wetland areas. [Reproduced by permission of Field Mark Publications.]

Wetlands are home to snowy egrets, which have long legs for wading in shallow water and sharp beaks for stabbing food. [Reproduced by permission of Field Mark Publications.]

A group of hippopotami dozing in a river. The name hippopotamus means "river horse" in Greek. [©1995 PhotoDisc.]

This oystercatcher is an example of a shorebird. These birds like to feed and nest along the banks of rivers and streams. [Reproduced by permission of JLM Visuals.]

An aerial view of the Amazon River running through a forest. [©1995 PhotoDisc.]

Crops along the Nile River. Yearly floods bring rich sediments to soil and make it ideal for farming. [Reproduced by permission of Archive Photos, Inc.]

A waterfall, such as the one pictured here, occurs when a river or stream falls over a cliff or erodes the channel to such an extent that a steep drop occurs. [©1995 PhotoDisc.]

up its food web. Along the seashore, as elsewhere, the food web consists of producers, consumers, and decomposers. These three types of organisms transfer energy within the seashore environment.

Phytoplankton are among the primary producers along the lower zone of the seashore. They produce organic (derived from living organisms) materials from inorganic chemicals and outside sources of energy, primarily the Sun. Other primary producers include seagrasses and plants that live in the upper zone.

Zooplankton and other animals are consumers. Animals that eat only plants are primary consumers. Secondary consumers eat the plant-eaters. They include zooplankton that eat other zooplankton. Tertiary consumers are the predators, like starfish and mice, that eat the second-order consumers. Some, such as mice and humans, are also omnivores, organisms that eat both plants and animals.

Decomposers, which feed on dead organic matter, include some insects and species of crabs. Bacteria also help in decomposition.

A serious threat to the seashore food web is the concentration of pollutants and dangerous organisms because they become trapped in sediments where the lower life forms feed. These life forms become food for higher life forms, and at each step in the food chain the pollutant becomes more concentrated. Finally, when humans eat contaminated sea animals, they are in danger of serious illness. The same process is true of diseases such as cholera, hepatitis, and typhoid, which can survive and accumulate in certain sea animals, and which are then passed on to people who eat those animals.

SPOTLIGHT ON SEASHORES

POINT REYES NATIONAL SEASHORE

Point Reyes National Seashore is part of a large peninsula (arm of land) extending into the Pacific Ocean. Its broad beaches are backed by tall cliffs and forested hills and valleys.

The shoreline along the California coast was carved by volcanoes and earthquakes, and earthquakes are still common. During the 1906 earthquake that struck San Francisco, the Point Reyes peninsula moved more than 16 feet (5 meters) to the northwest. Shifting sands, islands, and steep underwater cliffs are all found in the region.

POINT REYES NATIONAL SEASHORE

Location: Pacific coast of California, north of San Francisco

Area: 64,546 acres (25,818 hectares)

Wave action is strong and has carved the offshore rocks into rugged shapes. Rip currents and a pounding surf make swimming dangerous in some places. In sheltered areas, quiet bays and lagoons have formed, which are enclosed by sand dunes and grass-covered lowlands.

Sea stars, horseshoe crabs, and other invertebrates live along the shore, and Point Reyes is popular with seabirds, sea otters, and sea lions. Migrating gray whales can often be seen passing in the distance.

PADRE ISLAND NATIONAL SEASHORE

Padre Island National Seashore is located on a barrier island that stretches for about 100 miles (160 kilometers) along the coast of Texas in the Gulf of Mexico. Padre Island also borders the Laguna Madre, a shallow lagoon. The park consists of sandy beaches, and dunes as high as 40 feet (12 meters).

> **PADRE ISLAND NATIONAL SEASHORE**
>
> **Location:** Coast of Texas, along the Gulf of Mexico
>
> **Area:** 133,918 acres (53,567 hectares)

Many waterfowl winter on the island. Other birds include herons, terns, egrets, brown pelicans, and white pelicans. Since the mid-1980s, scientists have been bringing eggs from endangered sea turtles here to help rebuild the population.

The island is uninhabited by people and is the largest undeveloped beach remaining in the United States (exclusive of Alaska and Hawaii).

VIRGIN ISLANDS NATIONAL PARK

Virgin Islands National Park occupies about 60 percent of St. John, the smallest of the three main Virgin Islands, and takes in many tiny islands offshore. Many caves and grottoes can be found along the shoreline.

> **VIRGIN ISLANDS NATIONAL PARK**
>
> **Location:** Caribbean Sea, east of Puerto Rico
>
> **Area:** 15,150 acres (6,060 hectares)

The park's white sandy beaches are backed by steep mountains and valleys covered with tropical forests. The highest point of land is Bordeaux Peak at 1,277 feet (389 meters). Mangrove forests grow along the shore, while offshore a number of coral reefs support many tropical fish and other marine creatures such as sea urchins. The only native mammals that live within the park are bats.

Pre-Columbian Indians once lived here, and relics of their culture can often be found. For two centuries, the islands were also a base for pirates, and some people believe that buried treasure may still be hidden in the region.

ACADIA NATIONAL PARK

Acadia National Park includes parts of Mt. Desert Island, Isle au Haut, and Schoodic Peninsula, as well as several smaller islands. It is a rugged, rocky shoreline backed by mountains, and coniferous forests grow close to the water.

More than 10,000 years ago, the shoreline here was much further out than it is now. For that reason, the region is often called the "drowned" coast. Its cliffs were once inland mountains, and the park includes a fjord called Somes Sound. Some areas show the rubble left by glaciers.

Wave action of the North Atlantic is strong here, and huge blocks of granite have been dislodged from the cliffs and lie along the shore. Anemone Cave, a large cavern 82 feet (25 meters) deep, has been carved into the solid rock by wave action. In sheltered areas, small bays and coves have formed.

> **ACADIA NATIONAL PARK**
> **Location:** Atlantic coast of Maine
> **Area:** 41,634 acres (16,653 hectares)

The intertidal zone has many tidepools and is home to a host of creatures, including sea stars, periwinkles, anemones, sea urchins, and crabs. Seabirds, such as gulls, guillemots, and ducks are also common.

CAPE HATTERAS NATIONAL SEASHORE

Cape Hatteras National Seashore is a chain of low, narrow, barrier islands, including Bodie, Hatteras, and Ocracoke. In places, the islands are as much as 30 miles (48 kilometers) off the North Carolina coast. The seashore area has 70 miles (110 kilometers) of beaches, dunes, and salt marshes and is a wildlife refuge.

Cape Hatteras National Seashore is known for severe storms. Off the Cape, which is a promontory (arm of land) on Hatteras Island, cold northern currents meet the warmer Gulf Stream as it moves through the Atlantic. Their collision breeds stormy conditions. Diamond Shoals, a region of shallow water, has been called the "graveyard of the

> **CAPE HATTERAS NATIONAL SEASHORE**
> **Location:** Atlantic coast of North Carolina
> **Area:** 28,500 acres (11,400 hectares)

Atlantic" because so many ships have been wrecked here. Their skeletal remains can still be seen from the shore.

The Cape has had a lighthouse since 1798. The present lighthouse, which is the tallest in the United States at 208 feet (63 meters), has been in operation since 1870. However, years of severe weather, including hurricanes, threatened to destroy it and it was finally moved inland.

RHINE RIVER DELTA

The Rhine River Delta is a vast area connecting the Rhine River with the North Sea. Centuries ago, it was marshy and dotted with small islands.

However, storms often swept in from the North Sea, and the land was frequently flooded.

Drawn by the plentiful supply of fish, the Dutch began living on the islands in the Rhine River Delta around 450 B.C. They built hills of earth on which to live during flood times. Then, in the twelfth century, special walls called dikes were constructed to keep out the sea. Later, Dutch soldiers fighting in Middle Eastern lands learned a trick from the Arabs who used windmills to pump water for growing crops. The Dutch figured the same process could be used in reverse, to drain the land. Soon, the Dutch had built dikes around the marshy areas and used windmills to pump the water out into the ocean. In later years, steam-driven pumps replaced the windmills, and by 1850 more than 40,000 acres (99,009 hectares) near Amsterdam had been drained, allowing people to live there permanently.

> ### RHINE RIVER DELTA
> **Location:** North Atlantic coast of the Netherlands
>
> **Area:** Approximately 900 square miles (2,340 square kilometers)

In 1919, work began to reclaim a freshwater lake, called the Zuider Zee, that had been invaded by the sea, and by 1932 a dam was completed that allowed the seawater to drain. By 1937, the lake was fresh again. This lake is now called Ijsselmeer.

When a winter storm in 1953 damaged 400 miles (640 kilometers) of dikes along the coast, many people were killed when the land was flooded. This inspired the Dutch to build more dams across the delta so that such an accident could never be repeated. The project was finally finished in 1988. Protected by dikes, dams, and pumps, the Rhine Delta's fertile soil is now valuable farmland. About 40 percent of the Netherlands is below sea level during high tide. The lowest spot, 22 feet (6.7 meters) below sea level, lies to the northeast of Rotterdam.

FJORDS OF WESTERN NORWAY

Usually found in northern regions along mountainous coasts, fjords are valleys eroded by glaciers. When the glaciers retreated or melted, the ocean poured in, creating long, narrow, deep arms of water that project inland. The fjords of western Norway have high, steep walls—often rising from 3,000 to 5,000 feet (912 to 1,520 meters). The crowns of these surrounding cliffs may be heavily forested. Cascading waterfalls sometimes drop from great heights, adding to the spectacular scenery.

> ### FJORDS OF WESTERN NORWAY
> **Location:** North Atlantic coast of Norway
>
> **Individual Length:** As much as 114 miles (182 kilometers)

The depth of the fjords—the deepest along western Norway is the Sognefjord at 4,291 feet

(1304 meters)—depends upon how much erosion took place below sea level. Inland, some fjords end in glaciers, from which huge chunks of ice break off, creating icebergs.

Fjords are not always penetrated by cold ocean water and are usually ice free in the winter, except along the coast where the water may be more shallow and freezes more rapidly. Gales (high winds) are frequent on the western coast.

Inland fjord valleys are rich in vegetation, especially coniferous trees, lowland birches, and aspens. The coast attracts large numbers of food fishes, such as cod, herring, mackerel, and sprat. Woodcocks and migratory birds, as well as lemmings, hares, red foxes, and reindeer, are also common.

FOR MORE INFORMATION

BOOKS

Baines, John D. *Protecting the Oceans.* Conserving Our World. Milwaukee, WI: Raintree Steck-Vaughn, 1990.

Feltwell, John. *Seashore.* New York: DK Publishing, Inc., 1997.

Kaplan, Eugene H. *Southeastern and Caribbean Seashores.* NY: Houghton Mifflin, Company, 1999.

Lambert, David. *Seas and Oceans.* New View. Milwaukee, WI: Raintree Steck-Vaughn, 1994.

Massa, Renato. *Along the Coasts.* Orlando, FL: Raintree Steck-Vaughn Publishers, 1998.

ORGANIZATIONS

American Littoral Society
  Sandy Hook
  Highlands, NJ 07732
  Phone: 732-291-0055

American Oceans Campaign
  725 Arizona Ave., Ste. 102
  Santa Monica, CA 90401

Center for Environmental Education
  Center for Marine Conservation
  1725 De Sales St. NW, Suite 500
  Washington, DC 20036

Coast Alliance
  215 Pennsylvania Ave., SE, 3rd floor
  Washington, DC 20003
  Phone: 202-546-9554; Fax: 202-546-9609

Environmental Defense Fund
  257 Park Ave. South
  New York, NY 10010

Phone: 800-684-3322; Fax: 212-505-2375
Internet: http://www.edf.org

Environmental Network
4618 Henry Street
Pittsburgh, PA 15213
Internet: http://www.envirolink.org

Environmental Protection Agency
401 M Street, SW
Washington, DC 20460
Phone: 202-260-2090
Internet: http://www.epa.gov

Friends of the Earth
1025 Vermont Ave. NW, Ste. 300
Washington, DC 20003
Phone: 202-783-7400; Fax: 202-783-0444

Greenpeace USA
1436 U Street NW
Washington, DC 20009
Phone: 202-462-1177; Fax: 202-462-4507
Internet: http://www.greenpeaceusa.org

Sierra Club
85 2nd Street, 2nd fl.
San Francisco, CA 94105
Phone: 415-977-5500; Fax: 415-977-5799
Internet: http://www.sierraclub.org

World Wildlife Fund
1250 24th Street NW
Washington, DC 20037
Phone: 202-293-4800; Fax: 202-229-9211
Internet: http://www.wwf.org

WEBSITES

Note: Website addresses are frequently subject to change.

Discover Magazine: http://www.discover.com

Monterey Bay Aquarium: http://www.mbayaq.org

National Center for Atmospheric Research: http://www.dir.ucar.edu

National Geographic Society: http://www.nationalgeographic.com

National Oceanic and Atmospheric Administration: http://www.noaa.gov

Scientific American Magazine: http://www.scientificamerican.com

World Meteorological Organization: http://www.wmo.ch

BIBLIOGRAPHY

Amos, William H. *The Life of the Seashore.* New York: McGraw-Hill Book Company, 1966.

Barnhart, Diana and Vicki Leon. *Tidepools.* Parsippany, NJ: Silver Burdett Press, 1995.

"Beach and Coast." *Grolier Multimedia Encyclopedia.* Danbury, CT: Grolier, Inc., 1995.

Campbell, Andrew. *Seashore Life.* London: Newnes Books, 1983.

Cumming, David. *Coasts.* Habitats. Austin, TX: Raintree Steck-Vaughn, 1997.

Engel, Leonard. *The Sea.* Life Nature Library. New York: Time-Life Books, 1969.

Macquitty, Dr. Miranda. *Ocean.* Eyewitness Books. New York: Alfred A. Knopf, 1995.

McLeish, Ewan. *Oceans and Seas.* Habitats. Austin, TX: Raintree Steck-Vaughn Company, 1997.

"Ocean." *Encyclopaedia Britannica.* Chicago: Encyclopaedia Britannica, Inc., 1993.

Oceans. The Illustrated Library of the Earth. Emmaus, PA: Rodale Press, Inc., 1993.

Pernetta, John. *Atlas of the Oceans.* New York: Rand McNally, 1994.

Ricciuti, Edward R. *Ocean.* Biomes of the World. Tarrytown, NY: Benchmark Books, 1996.

Sayre, April Pulley. *Seashore.* New York: Twenty-first Century Books, 1996.

Wells, Susan. *The Illustrated World of Oceans.* New York: Simon and Schuster, 1991.

Whitfield, Philip, ed. *Atlas of Earth Mysteries.* Chicago: Rand McNally, 1990.

TUNDRA

In the northern lands close to the Arctic, and on the upper slopes of high mountains all over the world, a unique biome called the tundra is found. The word tundra comes from a Finnish word "tunturia," which means barren land. In this cold, dry, windy region where trees cannot grow, the often bare and rocky ground will support only hardy, low-growing plants, such as mosses, heaths, and lichens (LY-kens), which give it a greenish-brown color. During the brief spring and summer, flowers burst into bloom with the warmth of the Sun, dotting the landscape with color.

Tundra covers about 20 percent of the Earth's surface. Although almost all tundra is located in the Northern Hemisphere (the half of the Earth above the equator), small areas do exist in Antarctica in the Southern Hemisphere. Because Antarctica is much colder than the Arctic, however, the ground is usually covered by ice, and conditions are seldom right for tundra to form.

The tundra is not a biodiverse environment. (A biodiverse area is one that is able to support a wide variety of plants and animals.) Although only a few species of plants and animals live in the tundra, those few, such as lichens and mosquitoes, are found in great numbers.

The tundra may seem bleak and unfriendly, yet it can be a place of eerie beauty, especially in the Arctic during the winter. Here, winter nights last for weeks and are often lit up by the blue, red, and green colors of the Aurora Borealis, or northern lights. (The Aurora Borealis occurs when energy-charged particles from the Sun enter Earth's atmosphere and create flashes of light.) The Arctic is also called the Land of the Midnight Sun because, during the summer, the Sun never sets below the horizon and daylight lasts for 24 hours.

HOW TUNDRA DEVELOPS

Tundra forms primarily because of climate. In the Arctic, winters are long and cold, and summers are short and cool. This allows limited plant

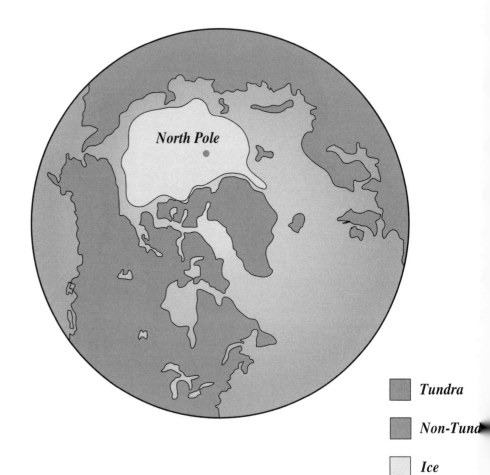

North Pole

Tundra

Non-Tund

Ice

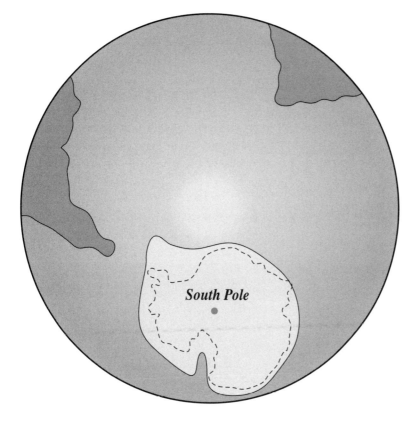

South Pole

growth. On high mountains, tundra forms when the location is right to produce the necessary climate.

The lack of soil in a tundra region may be due to erosion (wearing away) from wind and rain. During the Ice Ages more than 10,000 years ago, glaciers scraped away any soil, leaving only bare rock.

KINDS OF TUNDRAS

There are two types of tundra: Arctic and alpine. Arctic tundra is found near the Arctic Circle. Alpine tundra forms on mountaintops where the proper conditions are found.

ARCTIC TUNDRA

Several characteristics are typical of Arctic tundra. One is the polar climate, which has an average July temperature of not more than 50°F (10°C). Because Arctic tundra is far from the equator, sunlight hits the Earth here at an angle and must therefore pass through more atmosphere. This means the sunlight that reaches the soil contains less energy per square foot (square meter) than at the equator.

Another characteristic of the Arctic tundra is a deep layer of permanently frozen ground. Generally, less than about 18 inches (45 centimeters) of tundra soil thaws during the cool summer. Below that the ground remains frozen. Water from melting snow cannot drain into the frozen ground, and little evaporates in the cool summer air. As a result, the water becomes trapped on the surface.

Arctic tundra is found on all three northern continents, close to or above the Arctic Circle and near the Arctic Ocean. In Asia, Arctic tundra is found in that part of Russia known as Siberia. In Europe, it is found in northern Scandinavia, which includes the countries of Norway, Sweden, and Finland. In North America, Arctic tundra is found in northern Alaska and Canada. Some Arctic tundra is also located on islands, such as Greenland and Iceland.

STORMS ON THE SUN

The Aurora Borealis, or Northern Lights, occur when there are "storms" on the Sun. These storms shoot out streams of energy-charged particles called electrons. When the electrons enter the Earth's atmosphere and collide with each other, a flash of light occurs. When billions of these collisions occur at the same time, they create the magnificent Aurora Borealis.

ALPINE TUNDRA

Alpine tundra is found at the tops of mountains above the tree line, the point above which trees cannot grow. The tree line and the tundra are found at different elevations (heights) in different mountain ranges. The farther the mountains are from the equator, the lower the elevation needed for tundra to form.

Compared to Arctic tundra, alpine tundra gets more rain and its soil drains better because of the sloping terrain. It also gets more sunlight because it is found at lower latitudes (a distance north or south of the equator, measured in degrees) where day and night are more equal in length than in the Arctic. Usually, there is no lower layer of permanently frozen ground in alpine tundra.

Alpine tundra is found in the Rocky Mountains in North America, the Andes Mountains in South America, the Alps in Europe, and the Himalayas in Asia.

CLIMATE

Both Arctic and alpine tundras have very cold climates, although variations in temperature ranges may occur because of location. Arctic tundra located high above sea level (the average height of the sea) has a colder climate. If tundra is found near the coast, ocean currents can affect the temperature. For example, the North Atlantic Drift, a warm ocean current, warms the coast of northern Scandinavia. However, the coast of northeastern Canada is colder due to the influence of the Labrador Current, an icy current that mixes with the warmer waters. Arctic tundra is also windy, with winds ranging between 30 to 60 miles (48 to 97 kilometers) per hour.

Unlike Arctic tundra, alpine tundra has a more moderate climate that varies with latitude and altitude. The farther away from the equator, the colder

The Aurora Borealis, or northern lights, in the Arctic tundra. (Reproduced by permission of Corbis. Photograph by Dennis di Cicco.)

the temperature becomes. The higher the altitude, the colder and windier the climate. At 15,000 feet (4,500 meters) above sea level, the climate changes as much as if it were actually located 10° latitude farther north. High altitude also means that the atmosphere is thin, which results in little oxygen present in the air.

TEMPERATURE

The Arctic tundra is one of the driest and coldest biomes on Earth. January temperatures average from -4° to -22°F (-20° to -30°C). In Siberia, an extremely cold area in north central Asia, winter temperatures may drop to -40°F (-30°C) or lower. However, Scandinavian tundra is relatively warm, and the winter temperatures may average 18°F (-8°C). Average Arctic temperatures in summer range from 30° to 60°F (-10° to 16°C). Although temperatures above 90°F (32°C) have been recorded, normally they do not rise much above 50°F (10°C).

Antarctica's small tundra region is even colder, with an average annual inland temperature of -70°F (-56°C).

An illustration showing the vegetation zones in the alpine regions of the Himalayas.

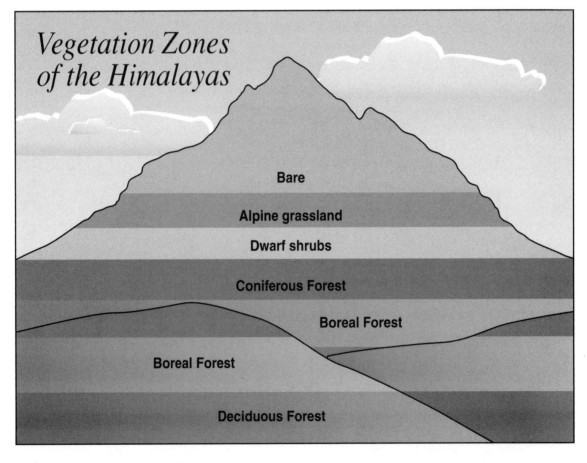

Vegetation Zones of the Himalayas

Bare

Alpine grassland

Dwarf shrubs

Coniferous Forest

Boreal Forest

Boreal Forest

Deciduous Forest

Although alpine tundra is also cold, temperatures are more moderate than in the Arctic. Elevation affects the temperature in an alpine tundra, and mountains have been described as "open windows letting the heat out." Scientists estimate that for every 300 feet (91 meters) in height, the temperature drops more than 1°F (almost 2°C).

Winter temperatures in an alpine tundra rarely fall below 0°F (-18°C) and summers are cool. The average annual temperature in the Peruvian mountains, for example, seldom falls below 50°F (10°C). On Alaskan mountains, temperatures in January average about 8°F (-13°C) and almost 47°F (8.2°C) in July.

PERCIPITATION

The Arctic tundra is similar to a cold desert in that it receives little precipitation (rain, snow, or sleet)—usually less than 10 inches (250 millimeters) annually. Most precipitation falls as snow in winter.

A photo of the Alps in Europe. Like other mountains, alpine tundra is found in these mountains. (Reproduced by permission of Susan Rock.)

Alpine tundra receives more rain than Arctic tundra, but the water runs rapidly off the mountain slope, leaving any dry soil to blow away in the wind.

GEOGRAPHY OF TUNDRAS

The excessively cold temperatures also affect the geography of the tundra.

LANDFORMS

In the Arctic tundra, the water trapped in the top layer of soil does strange things to the landscape. The water freezes each winter, and the ice melts each summer. When water freezes, it expands (takes up more space). When ice melts, it contracts (shrinks). This yearly expansion and contraction cracks and breaks rocks and creates hills, valleys, and other physical features.

A pingo is a small circular or oval hill formed when a pool of water under the ground freezes and forces the soil up and out. The hill may grow a few inches taller every year. Some pingos are as high as 300 feet (90 meters) and more than half a mile (800 meters) wide. Stone circles are formed by piles of rocks that have been moved into a more or less circular shape by the expansion of freezing water.

Polygons are cracks in the ground that take on geometric shapes because of the freezing and thawing action. If the soil is rocky, the rocks can be pushed up through the cracks, making the geometric shapes even more distinct.

Irregularly shaped ridges, called hummocks or hammocks, are formed when large blocks of ice meet and one slides over the top of the other. When the ice melts, the ground is uneven. Their heights range from 65 to 100 feet (20 to 30 meters).

Areas of bare, rock-covered ground on the alpine tundra are called fell-fields. They are often formed when rock and soil slide down a slope.

ELEVATION

Most of the low, rolling plains of the Arctic tundra are located about 1,000 feet (300 meters) above sea level, whereas alpine tundra is found high on mountains above the tree line. In northern latitudes, close to the Arctic

---

ARCTIC HEAT WAVE

Sometimes the weather can become quite warm during summer on the Canadian tundra, warm enough for people to swim in the Arctic Ocean. In 1989 the temperature rose above 90°F (32°C) at Coppermine in the Northwest Territories. These temperature increases are caused by warmer water brought into the Arctic Ocean by northward-moving currents, such as the Irminger current, a branch of the Gulf Stream, which is a warm ocean current that begins in the Gulf of Mexico.

A DANGEROUS CHILL

The wind chill factor is the combination of cold temperature and strong wind that makes the effect of cold weather on the body worse than just the temperature would indicate. For example, if the temperature is 20°F (-6°C) and a 20 mile- (32-kilometer-) per-hour wind is blowing, the temperature will feel like -10°F (-23°C). In the Arctic, wind chill factors are so severe that bare flesh can freeze within 30 seconds.

Circle, arctic tundra may begin at about 4,000 feet (1,200 meters) above sea level. In more southern mountains, alpine tundra usually begins around 8,000 to 10,000 feet (2,438 to 3,048 meters) above sea level.

SOIL

The soil in the Arctic tundra has two layers. The surface layer is called the active layer because it freezes in winter and thaws when the weather warms up. This layer is shallow, its depth ranging from about 10 inches to 3 feet (25 to 100 centimeters.) About 15 percent of the active layer is well drained because it is located on stony or gravelly material, on slopes, and in elevated areas. The remaining 85 percent of the active layer, however, is usually poorly drained and remains wet.

The lower, or inactive, layer of soil stays frozen throughout the year and is called permafrost. Permafrost prevents the water captured in the active layer of soil from draining away. Made of such materials as gravel, bedrock, clay, or silt, permafrost reaches depths of 300 to almost 2,000 feet (90 to 609 meters). In Russia on the Taimyr Peninsula, permafrost goes very deep, reaching 1,968 feet (600 meters), whereas permafrost on tundra near Barrow, Alaska, descends to only 984 feet (300 meters).

Tundra soil is generally poor in nutrients, such as nitrogen and phosphorous. In some areas, however, where animal droppings are plentiful and fertilize the soil, vegetation is lush. Near the southern edge of the Arctic tundra, for example, the soil can be boggy. Bog soil contains little oxygen, is acidic, and is low in nutrients and minerals.

> ### TREASURES IN THE ICE
>
> The remains of plants and animals, millions of years old, have been discovered in the tundra permafrost. Tree stumps found on the Canadian tundra could still be burned for fuel.
>
> The bodies of many mammoths have been perfectly preserved in the Siberian tundra. These large animals, now extinct, are ancestors of the modern-day elephant. In the early 1900s, one was discovered with its head sticking out of a bank of the Berezovka River. Although wolves had eaten part of his head, its tongue and part of its mouth were still preserved. In its mouth, between its teeth, were the remains of the sedge and buttercups it had been eating. Both of these plants still grow on the tundra today.

Although alpine tundra rarely has a permafrost layer, the soil does freeze and thaw, as in the Arctic. When permafrost does occur in alpine tundra, it is at higher elevations where temperatures are colder and in areas where mud, rock, and snow slides are common.

Unlike arctic soil, alpine soil is also stratified (arranged in layers). Because of the sloping terrain, alpine soil also has good drainage. However, fierce winds may dry it out and blow it away. Some alpine regions are covered with material that is so weathered and thin it cannot even be classified as soil.

WATER RESOURCES

Melting snow in the Arctic tundra has nowhere to go since it cannot sink into the ground, and temperatures never get warm enough for it to evaporate. As a result, in summer the tundra is covered with marshes, lakes, bogs, and streams. Marshes are a type of wetland characterized by poorly drained soil and plant life dominated by nonwoody plants. Bogs are a type of wetland that has wet, spongy, acidic soil, called peat.

Thermokarst, or thaw lakes, are shallow bodies of water unique to the Arctic tundra and formed by melting ground ice. Permanent rivers also flow into the tundra, like the Mackenzie and Yukon rivers in Alaska, and the Lena, Ob, and Yenisei in Siberia. These rivers are partially or completely covered by ice for about six months of the year.

In the alpine tundra, water from melting snow and glaciers usually runs off the slopes. In areas where depressions in the ground occur, ponds and marshes form. Mountain streams, which flow during the five warm months of the year, are formed by surface runoff and from springs.

PLANT LIFE

Permafrost and the yearly freezing and thawing, break up plant roots and make it impossible for trees and other tall plants to survive on the Arctic tundra. Plant growth in the alpine tundra is also affected by freezing and thawing. Because alpine tundra occurs on mountains, the sloping terrain, exposure to more light, and the lesser amount of moisture available also affect the kinds of plants that grow here.

Despite the harsh conditions, about 1,700 kinds of plants grow on the tundra, including sedges, reindeer mosses, liverworts, and grasses. The 400 varieties of flowers found here add a wide range of colors.

ALGAE, FUNGI, AND LICHENS

It is generally recognized that algae (AL-jee), fungi (FUHN-jee), and lichens do not fit neatly into the plant category. In this chapter, however, these special organisms will be discussed as if they, too, were plants.

Algae Algae play an important role in the tundra. Most are single-celled organisms; only a few are multicellular. Many species have the ability to make their own food by means of photosynthesis (foh-toh-SIHN-thuh-sihs; the process by which plants use the energy from sunlight to change water and carbon dioxide from the air into sugars and starches). Other species of algae absorb nutrients from their surroundings. In the Arctic, algae are found in the surrounding ocean and in wetland areas. On the alpine tundra, algae grow on unmelted snow. These algae have a red pigment, which collects heat from the Sun and keeps them warm enough to grow.

Algae may reproduce in one of three ways. Some split into two or more parts, with each part becoming a new, separate plant. Others form spores (single cells that have the ability to grow into a new organism). A few, however, reproduce sexually, during which male and female cells unite to create a new plant.

Fungi Fungi are plantlike organisms that cannot make their own food by means of photosynthesis; instead they grow on decaying organic (derived from living organisms) matter or live as parasites on a host. Fungi are decomposer organisms that, together with bacteria, are responsible for the decay and decomposition of organic matter. One of the most important roles of fungi in the tundra is in the formation of lichens.

Fungi form spores to reproduce. These spores are carried from one location to another on the air or by animals.

Lichens Lichens are the most common tundra plant, with about 2,500 species growing in Arctic and alpine regions. Lichens are actually combinations of algae and fungi living in cooperation. The fungi surround the algae cells. The algae obtain food for themselves and the fungi by means of photosynthesis. It is not known if the fungi aid the algae organisms, although they may provide them with protection and moisture. In harsher climates, lichens are often the only vegetation to survive. Because they have no root system, they can grow on bare rock, adding beauty to the tundra with their colors of orange, red, green, white, black, and gray.

Despite the short growing season, lichens thrive beautifully in their harsh environment. Although they freeze in winter, in spring they continue to grow. Lichens often live for hundreds of years although growth is slow.

Like algae, lichens can also reproduce in several ways. If a spore from a fungus lands near an alga these two different plants can join together to form a new lichen. Lichens can also reproduce by means of soredia (algal cells surrounded by a few strands of fungus). When soredia break off and are carried away by wind or water they form new lichens wherever they land.

Reindeer lichen is one of the most common tundra plants. It provides a key source of food for Arctic plant-eating animals, like the caribou and reindeer. Another common type of lichen, called the British soldier, has a tall stalk with a red cap on top, which makes it resemble its namesake.

GREEN PLANTS

Most green plants need several basic things to grow: light, air, water, warmth, and nutrients. Light, air, and water provide almost all of their needs. The remaining nutrients—primarily nitrogen, phosphorus, and potassium—are obtained from the soil. While water is plentiful for tundra plants, warmth, which brings on the growing season, is available for only a short period of time.

Plants in the tundra grow low to the ground. This helps them to stay warm and protected from the high winds. Since the tundra receives little precipitation, many plants have almost invisible hairs on their leaves, stems, and flowers that reduce moisture loss. Their roots form a dense mat just under the surface of the ground, which enables them to quickly draw moisture from melting snow and store it in their leaves. Many plants grow in a tight clump, which helps traps heat. Tussocks, thick clumps of plants, such as cotton grass, that grow to about 1 foot (30 centimeters) in height are found in marshy areas.

NATURE'S HOTHOUSES

Since the tundra growing season is so short, some plants get a head start by making use of "hothouses" formed by the Sun. Because darker colors absorb more heat, the Sun melts some snow that is close to the dark soil. This forms small caves in the snow. The floor is the soil and the roof is a dome of snow that remains frozen. Poppies and saxifrages grow well in these miniature hothouses because the air inside is warmer than the outside air.

Some alpine plants, however, have deeper roots which help prevent soil erosion. Many alpine plants have red leaves because of a coloring matter called anthocyanin. This special pigment absorbs the Sun's warmth and protects the plants from the dangers of ultraviolet radiation (a form of harmful light within the Sun's rays). Because of the high elevation, twice as much ultraviolet radiation reaches alpine tundra than regions at sea level.

Growing season The growing season on the tundra is short, lasting from six to ten weeks in June, July, and August. In the Arctic, during those weeks the Sun shines almost all day. As a result, plants must make the most of the opportunity for growth.

Green plants may be annuals or perennials. Annuals live only one year or one growing season. Perennials live at least two years or two growing seasons. The above ground portion often appears to die in winter, while the roots remain dormant underground. The plant then returns to "life" in the spring when the weather becomes warmer. Most tundra plants are perennials that are able to flower quickly and take advantage of the short growing season.

Reproduction Perennials reproduce and spread by forming seeds. Some plants are self-pollinating, which means that the male and female reproductive cells come from the same plant. Most perennials, however, need pollen from another plant. This process is aided by insects attracted by the colors of the flowers and that travel from plant to plant transferring the pollen from the male reproductive part of a plant, called the stamen, to the female reproductive part of another plant, called the pistil. Strong winds help scatter the fertilized seeds.

Because the growing season is short, some plants in the more northern, colder Arctic tundra reproduce by budding and division rather than by flowering. In budding, a new plant simply develops from any part of the parent. In division, a piece of the parent plant breaks off and develops into a new plant. Forbs, a category of flowering plants, reproduce in this way. Mountain sorrel reproduces through rhizomes (RY-zohmz), rootlike stems that spread out under the soil and form new plants.

Common green plants Typical flowering plants on Arctic tundra include Arctic lupine, yellow poppy, saxifrage, Arctic campion, Lapland rhododendron, buttercup, campanula, and barberry. The Labrador tea, a hardy evergreen, is very common. One category of perennial, called forbs, lie dormant all winter with their growth buds protected underground.

The rosette plant is well-designed to survive in the harsh Arctic weather. This type of plant forms rings of leaves around a central growth bud. The leaves protect the fragile bud and help to trap insulating snow in winter and dew during the dry growing season.

Examples of alpine tundra plants include sorrel, saxifrage buttercup, plane leaf, tufted hair grass, alpine bluegrass, and alpine sandwort. The alpine azalea is a member of the heath family called a cushion plant. These cushion plants grow in groups, tightly clumped together, so the plants on the outer edge can protect the ones in the middle. At the lower edges of the alpine tundra grow Krummholz, dwarf trees that are kept small by the cold, icy, windy environment.

> ### THE COZY FLOWER
> Some tundra insects find warmth and shelter by hiding inside flowers. Inside a buttercup, for example, the temperature can be 40°F (4°C) warmer than the outside air. During the day, these flowers lean toward the Sun, following its path to get as much heat as possible. For that reason, they are called "heliotropes," which means "to turn toward the Sun."

The small area of tundra located on the Antarctic Peninsula is home to the only two flowering plants found on that continent, pearlwort and hair grass.

ENDANGERED SPECIES

Tundra plants are very fragile. In the Arctic, traffic during the construction of the Alaska oil pipeline, begun in the 1970s, damaged the tundra, which did not recover until long after the construction had ended. In the alpine tundra, beautiful plants are endangered because people pick them or step on them. Because their growing season is so short, they are not easily replaced.

The Penland alpine fen mustard, a small perennial with white flowers found on the Rocky Mountain tundra in Colorado, grows on wetlands fed by melting snowfields. This little plant is endangered because mining and off-road vehicles, which leave tracks in the soil, divert the flow of water away from its habitat.

ANIMAL LIFE

The tundra is a permanent home to only a few species of animals because of its harsh environment. Birds, caribou, and red deer, for example, spend only the summers there. The Antarctic tundra has the fewest animals of all.

For tundra animals, size is an important factor in preventing heat loss. When an animal's appendages (arms, legs, tails, ears) are small, they lose less heat. The Arctic fox, for example, has small ears, short legs, and a short tail. This means there is less area from which body heat can escape.

> ## MOSQUITO ANTIFREEZE
>
> Mosquitoes can keep themselves from freezing in winter in the same way people keep their cars from freezing—with antifreeze. Some mosquitoes replace the water in their bodies with a chemical called glycerol. Their bodies are able to manufacture the glycerol from other substances such as fat. With this protection, they can spend the winter under the snow and live to bite again the following summer.

MICROORGANISMS

A microorganism is a tiny animal, such as protozoa or bacteria, that cannot be seen with the human eye. Bacteria live in the active layer of the tundra soil and help decompose dead plants and animals.

INVERTEBRATES

Invertebrates are animals that do not have a backbone. Clams, mussels, snails, crabs, and shrimp are invertebrates found in marshy areas. However, the primary invertebrate population of the tundra is made up of a few species of insects that are present in large numbers. There are more mosquitoes on the Arctic tundra, for example, than anywhere else on Earth, because the wet summers provide perfect breeding conditions. As many as 1,000,000 mosquitoes can be found in an area 1 yard square (0.8 meter square). Springtails are the dominant organism of tundra soil. Populations can reach in excess of several billion per acre (0.405 hectare).

Arctic insects are darker in color than insects elsewhere, which is better for absorbing heat from the Sun. They also have more hair, although it is not know why this is the case. Many tundra insects do not have wings. Some scientists believe this helps them conserve energy. Others believe it is because they would not be able to fly in the strong winds. The springtail, for example, has a springlike appendage on it's stomach. It hits the ground with this spring and then bounces to a new location, much like a person on a pogo stick. Those that can fly, stay close to the ground so they don't get blown too far away.

Food Some insects eat plants and some eat other insects or animals. Male mosquitoes, for example, feed on plant juices, while females feed on the blood of animals and humans. During the summer, plenty of water is available in the Arctic tundra because the surface water cannot sink into the frozen ground. In the alpine tundra, surface water and streams are sources of water.

Reproduction The first part of an invertebrate's life cycle is spent as an egg. The second stage is the larva, the immature or young stage of a developing insect. This second part of the life cycle may actually be divided into several steps between which the developing insect grows larger and sheds an outer skin casing. Some larvae store fat in their bodies and do not need to seek food. During the third, or pupal, stage, the insect's casing does offer as much protection as an egg. Finally, in the last stage, the adult emerges. These stages can occur quickly or take up to several years. In the colder tundra regions the process is slow. The midge, for example, takes two years to complete its life cycle in the tundra, but only six months in warmer climates.

In order to mate, some wingless, female insects in the tundra emit a scent into the air that attracts the males, which can fly. In species where both males and females fly, the insects swarm in great numbers. Males are able to recognize females by the vibrations given off by the beating of their wings. In some species of Arctic insects, like the caddis fly and midge, males are few. Females carry all the necessary genetic information and reproduce without having to mate with a male. The eggs are stimulated to develop by chemical means rather than by male sperm.

Common invertebrates Two common tundra insects are the aphid and the midge. Aphids, also called plant lice, feed on the sap from willows, saxifrages, sedges, and other plants. Most aphids are wingless females. During the warmer weather, the females produce live young, which are also all females, without mating. In the fall, they give birth to a generation that includes both males and females. After mating, the females of this generation lay eggs that hatch in the spring to start new colonies.

> ## BUMBLEBEE HEATING SYSTEM
> Arctic bumblebees are the only insects able to produce their own heat. They vibrate their wing muscles so fast that they actually warm their bodies. When enough bees get together, this combined body heat is able to warm their nests to 86°F (30°C).

Midges are small insects with two sets of wings. One set is for flying and the other for balance. Unlike the mosquito, midges do not bite. There are so many of them, however, that their mating swarms are said to look like tornado clouds and can obstruct the vision of any animal they encounter. Midge larvae live in watery areas and feed on algae and dead matter. Midges can grow very old, some living as long as six years.

While the Arctic tundra is infamous for its mosquito population, the alpine tundra is known for butterflies and biting flies. In the small area of tundra in Antarctica, spider-like mites and wingless springtails are found. The largest animal living here is the midge, which is half an inch (1.27 centimeters) long.

AMPHIBIANS

Amphibians, such as salamanders and frogs are vertebrates (animals with a backbone) that live at least part of their lives in water. Most amphibians are found in warm, moist, freshwater environments and in temperate zones (areas in which temperatures are seldom extreme). Only a few amphibians live in the Arctic tundra, while none are usually found in the alpine tundra.

Food Amphibians use their long tongues to capture their prey. Even though they have teeth, they do not chew but swallow their food whole. As larvae, amphibians eat mostly plants, such as algae, or tiny water animals. Adults eat insects, worms, and other amphibians or small animals.

Reproduction Mating and egg-laying for most amphibians takes place in water. Male sperm are deposited in the water and must swim to the eggs, which are deposited in a jelly-like substance. Here the sperm penetrate the eggs. As the young develop into larvae and young adults, they often have gills and require a watery habitat. Once they mature, though, they develop lungs and can live on land.

Some amphibian females carry their eggs inside their bodies until they hatch. Certain species, however, lay eggs and protect them until they hatch. Others lay eggs and abandon them.

Most amphibians reach maturity at three or four years. They breed for the first time about one year after they become adults.

Common amphibians The Siberian salamander is the only member of the salamander family found in the tundra. This hardy creature spends winter frozen in the permafrost in Russia and has been found alive at depths of almost 50 feet (14 meters).

Aphids, like the ones pictured here, are common insects found in tundra regions. (Reproduced by permission of Corbis. Photograph by George Lepp.)

The Hudson Bay toad lives in Arctic tundra in North America, and the wood frog, which is common in Alaska, is sometimes found on the tundra there.

REPTILES

Reptiles are cold-blooded vertebrates, such as lizards and snakes. Cold-blooded animals are unable to regulate internal body temperature and depend on their environment for warmth. They are usually most active when the weather is warm. However, no reptiles do well in extreme temperatures, either hot or cold. Many hibernate (remain inactive) during the winter.

Food Most reptiles are carnivorous (meat eating). Snakes consume their prey whole—and often alive—without chewing, taking an hour or more to swallow a large victim. Many have fangs that are curved backward so their prey cannot escape.

Reproduction Most reptiles reproduce sexually with the male depositing sperm in the body of the female, and their young come from eggs. Reptiles in the tundra, however, give birth to live young that have developed inside the mother's body, which protects them from the cold.

Common reptiles The viviparous lizard in Europe and the garter snake in North America live on the tundra borders. Only one reptile, the European viper, lives north of the Arctic Circle on the Scandinavian tundra.

The European viper is a highly venomous (poisonous) snake found only in Europe, Asia, and Africa. It grows to an average length of 19 to 23 inches (50 to 60 centimeters), but some may attain 31 inches (80 centimeters). It dines on small vertebrates, such as rats or mice, and sometimes on birds. The viper paralyzes or kills its prey by injecting it with venom.

This viper tolerates cold better than any other member of the viper family. It is still able to move at temperatures as low as 37°F (3°C) and cannot thrive in weather much warmer than 93°F (34°C). During the coldest parts of the winter, it hibernates, hiding from 6 to 19 inches (15 to 50 centimeters) underground in areas that are moist, like rock crevices or rodent dens.

FISH

Fish are cold-blooded vertebrates that live in water. They have fins, which they use to swim with, and breathe through gills. Some tundra fish, like the Arctic char, spend their entire lives in tundra lakes and streams. Others, like the salmon, live in the ocean, returning to the tundra streams to breed.

Food Some fish eat plants while others prefer insects, worms, and crustaceans. Many fish are predators (hunters), the larger ones eating the smaller ones.

Reproduction Fish reproduce by laying eggs. The female releases the eggs and the male releases sperm and fertilization takes place in the water. These eggs can cling to rocks or float on the water's surface. Some species dig depressions in the ground under the water and deposit the eggs there.

Salmon spend from three to five years at sea before they return to the same stream where they were born to spawn (reproduce). When spawning is done, they die.

Common fish The Arctic char, a relative of the salmon, is found in both the sea and in polar lakes in Arctic tundra. Char are dark colored with light spots and blend in with their environments. Their size ranges from 2 to 10 pounds (0.9 to 4.5 kilograms). Those that live in the sea swim up the rivers of the Arctic tundra to freshwater lakes where they lay their eggs. Graylings, another relative of the salmon, live in tundra streams. Pike and trout also spend some time in the tundra.

There are no species of fish that live in the alpine tundra.

Two Arctic char caught in the Senja lake in Norway. (Reproduced by permission of Corbis. Photograph by Rick Price.)

BIRDS

Birds are vertebrates. Hundreds of species of birds visit the Arctic and alpine tundra, but few stay all year long. However, the snowy owl, the raven, the willow ptarmigan (TAHR-mah-guhn), and the rock ptarmigan are perma-

nent residents of the Arctic. The only species of bird found in both the Arctic and alpine areas is the water pipit.

Many species of birds migrate to the Arctic tundra to breed. They fly south in winter and return to the tundra in summer with their mates to build nests and lay their eggs.

Food The marshy areas of the tundra provide plenty of water for all animals, and many birds feed on water insects. Plant material, shellfish, and carrion are other sources of food. The golden eagle feeds on small mammals, usually rabbits. It will also eat other birds and carrion.

Reproduction All birds reproduce by laying eggs. Usually, the male birds must attract the attention of the females by singing, displaying feathers, or stomping their feet. After they mate and the female lays her eggs, she sits on them to keep them warm until they hatch. Two species of Arctic swimming birds, the red phalaropes and the northern phalaropes, reverse these roles. The female shows off to attract the attention of the male. After they mate, she usually lays four eggs and the male sits on them until they hatch and then he cares for the young.

Some birds, like the least sandpiper, take advantage of the short breeding season by rotating parental duties. After the female lays the eggs, the male mates with one other female and then returns to the nest of the first female to take care of the young. This frees the female to mate again with another male, ensuring that enough young birds will be born to survive the cold weather.

Many birds, like the ruddy turnstone, shelter along rocky coasts and shores. Their nests are hidden in depressions in the ground, called scrapes, that they have lined with tundra grasses.

Common birds Water birds that migrate to the Arctic tundra include geese, ducks, and even a few swans. The Arctic tern makes the longest journey of all, flying from Antarctica in a round trip of at least 25,000 miles (40,000 kilometers).

Birds that summer on the Arctic tundra include gulls, plovers, redpolls, buntings, loons, warblers, red phalaropes, and skuas. Some of these birds eat insects, which are plentiful during the summer. The skua and the gyrfalcon are predators that eat other birds and small animals.

The Alpine tundra is also home to many birds. Some species of hawks, eagles, and falcons live above the tree line and eat small mammals, such as rodents and hares, or smaller birds. Wall-creepers eat insects and may move lower down the mountain in winter. Alpine cloughs, wall-creepers, and accentors stay close to the ground to avoid the winds. Some finches remain at the higher elevations all year.

The ptarmigan is a brownish-colored grouse with feathered legs and feet. It is one of the few birds that spend nearly all year on the tundra, both in Arctic and alpine regions. Its feathers grow very dense and turn white during the winter. Sometimes ptarmigans sleep in the snow in order to preserve body heat, which is more easily lost out in the open. The ptarmigan is a plant-eater and, in very severe winter weather, travels below the tree line in search of food.

MAMMALS

Mammals are warm-blooded vertebrates (having a backbone). They maintain a consistent body temperature, are covered with at least some hair, and bear live young.

In a barren land such as the tundra, animals that live here all year long must be very clever in seeking shelter from both predators and the cold. Mammals adapt in several ways. Many, like the Arctic fox, grow a thick, insulating cover of fur, which not only keeps them warm, but also helps to camouflage them. Their coats turn white in winter and brown in summer so that they blend in with the environment and are hard for predators to see. Most accumulate deposits of fat under their skins, which help to insulate them and provide a source of nourishment when food is scarce. Some mammals, like the Alaskan marmot and the Arctic ground squirrel, wait out the cold by hibernating until the weather turns warm.

Alpine mammals must adapt to mountain living as well as the cold. Ibexes, for example, have soft pads on their feet that act like suction cups to

A willow ptarmigan in Canada. Ptarmigans are one of the few birds that spend nearly all year on the tundra. (Reproduced by permission of J L M Visuals.)

help them grip the steep rocks. Since there is less oxygen high in the mountains, the yak and mountain goat have developed large hearts and lungs and have more red blood cells (which absorb oxygen) than would be found in similar animals living closer to sea level. These features help their bodies use oxygen more effectively. Alpine mammals often seek shelter in forests below the tundra or in rock caves underground. However, yaks and musk oxen can withstand the worst cold the tundra has to offer.

Food Mammals are either meat-eaters or plant-eaters. Musk oxen and caribou are among the tundra's plant-eaters. Reindeer moss, a type of lichen, is the bulk of the caribou's diet. Other vegetarians include lemmings, hares, and squirrels.

The barren-ground grizzly and the polar bear are primarily meat-eaters, dining on seals, birds, and fish. The polar bear supplements its diet with seaweed and grass. Wolves eat hares, musk oxen, and caribou. Foxes eat lemmings, stoats, ptarmigans, and hares.

Reproduction Mammals give birth to live young that have developed inside the mother's body. Some mammals, like lemmings, are helpless at birth, while others, like caribou, are able to walk and even run almost immediately. Some are even born with fur, and with their eyes open.

Polar bears give birth in dens under the snow in the Arctic tundra. Caribou calves are born in the open as the animals migrate north in spring to their tundra feeding grounds. Musk oxen are also born on open tundra.

> **THE MONSTER OF THE MOUNTAINS**
> The Sherpa people of Nepal believe they share their mountain home with a creature even larger than the yak—the Yeti, or abominable snowman. A yak can weigh up to 2,200 pounds (1,000 kilograms), which means a Yeti must be huge! Although many footprints have been found and a number of Sherpa claim to have seen one, no Yeti, dead or alive, has ever been proven to exist.

Common mammals About 48 species of land mammals live on the tundra. Large Arctic mammals include musk oxen, caribou, barren-ground grizzly bears, wolves, and polar bears. Smaller Arctic mammals include hares, lemmings, and squirrels. Large alpine mammals include mountain goats, wild sheep, ibexes, red deer, and snow leopards. Pikas, marmots, chinchillas, hares, and viscachas are among the smaller alpine mammals.

CARIBOU Caribou are migratory herd animals that spend their summers in the Arctic and their winters in the forest on the edge of the tundra. Their winter coat is so long and thick that it protects them from freezing. Their hooves are wide and flat for easy movement over the snow and ice.

The name "caribou" is a Native American word that means "wandering one." Caribou migrate farther than any other land animal, some traveling over 1,200 miles (2,000 kilometers) each year. They are found in Alaska, the Yukon, and the Northwest Territories of Canada.

In summer, caribou feed on birch leaves and grasses. In the winter they feed chiefly on lichens, a staple in their diet, and also on what twigs and tree buds remain.

Caribou give birth to their young on the Arctic tundra, usually in the same area that the herd has used for many years. After only three days, a calf is strong enough to travel with the adults.

PIKA The pika is a small, short-legged animal that resembles a cross between a guinea pig and a rabbit. An adult pika weighs about 5 ounces (140 grams), has rounded ears, a stocky body, and almost no tail. A pika's claws, which are curved and quite sharp, help it climb on rocks, and the pads of their feet have thick fur to ensure stable footing when they leap from rock to rock. Protected against the cold by its thick fur, a pika can remain active all winter.

Pikas have two sets of upper teeth, one behind the other. These teeth are sharp and used for tearing the plants they eat for food. They produce two different types of droppings. One is a special, jelly-like pellet that the animal eats to gain extra nutrients that may not have been absorbed the first time the food went through its digestive tract. The second is a normal solid pellet of unneeded waste matter.

Pikas are found in rocky areas, usually at high altitudes. Colonies are formed around large boulders or rock slides where vegetation can be found nearby. They collect plant materials in the summer and make "haystacks," which are their winter food stores. These haystacks consist of their chief food-plant, the avens, a member of the rose family.

Male and female pikas mark an area in which they will live all year. They usually have two to six young in a litter and produce two litters every summer. Females generally do not leave their territories, but the males roam in search of food.

Pikas are most common in Asia, especially northeast Siberia, but a few species are found in the Arctic and on alpine tundra in Alaska.

MUSK OXEN When winter comes to the Arctic tundra, most animals head for shelter. Some go south while others go underground. Musk oxen do neither. Their thick coats keep them warm in the

LEMMINGS ON THE MARCH

Lemmings are small plant-eating mammals related to field mice. During winter they live in the shelter of burrows in the snow, feeding on seeds and plants. In years when there is a good food supply, lemmings reproduce rapidly. One female may give birth four or five times a year and have six to ten babies at a time. After three to five good years, the tundra is swarming with lemmings. Before long, however, there is not enough food for all of them, and something strange happens. Suddenly, small groups of lemmings begin to run north.

As the lemmings run, they are joined by more and more of their fellow creatures until thousands are on the march. They cross mountains and rivers, never stopping to eat. Many starve to death along the way and many more are eaten by predators.

When the survivors reach the Arctic Ocean they jump in the water. Some scientists say it is an attempt to cross it, some say it is to get food. Either way, the ocean is just too large and they all drown. The tundra now seems empty, but the few lemmings that stayed behind will soon repopulate it.

worst Arctic weather, and they remain on the open tundra all winter long, foraging under the snow for plants to eat. In spring, musk oxen shed their warm undercoat, called qiviut. This hair is highly prized by local women who gather it as it is shed each year to knit it into garments, such as scarves, to be worn or sold.

WOLVES Wolves live together in family-like packs of five to twenty members on both Arctic and alpine tundra. They are sociable animals, having affectionate feelings for each other, which helps the pack remain together. Each pack is led by a male and female who are called the alpha (lead) pair. These two wolves produce the yearly litter, and other members of the pack help take care of the pups. Pack members may babysit so the mother can join in a hunt. When a kill is made, a wolf can store partially digested meat in its stomach. This is then regurgitated (vomited) and is fed to the pups or is shared with a nursing mother.

ENDANGERED SPECIES

Overhunting and overfishing are two of the most common reasons that animals in the tundra are endangered. Their habitats are also being disrupted as more people move into the area. Some animals, like the snow leopard and the chinchilla, are now rarely seen in the wild but are raised in captivity in zoos or on fur farms.

A pika on a rock. Pikas are mammals found in the rocky, tundra regions of Asia, especially Siberia. (Reproduced by permission of Corbis. Photograph by D. Robert Franz.)

The Eskimo curlew is a dark-colored shorebird with a long, downward-curving bill. At one time, the nesting grounds of the curlew were on the Arctic tundra of Alaska and Canada and on the coast of the Chukotka Peninsula in Siberia. The bird is endangered because of unrestricted hunting. Its habitats have also been destroyed as land is cultivated for farming and used for grazing herds. Scientists are no longer sure where its current breeding grounds are, and it is probably close to extinction.

Caribou in Greenland and the Peary Caribou in the Canadian Arctic are two groups whose numbers are low. Caribou are affected by predators such as wolves, overhunting by humans, and changes in their habitats such as natural gas pipelines that cut across their migration routes.

THE WARMEST COAT ON THE TUNDRA

The chinchilla is a small South American rodent that lives high in the Andes Mountains. It has one of the warmest coats found in nature, a beautiful blue-gray fur. Usually, a mammal grows only one hair per pore, but chinchillas have as many as 60 hairs growing in each pore. As a result, their fur is extremely thick and soft, keeping the chinchilla quite comfortable in the cold, even during the freezing alpine nights.

HUMAN LIFE

Few people live on the Arctic tundra but those who do are spread all around the polar region. Most people who live in areas of alpine tundra make their homes below the tree line. In some areas, however, people live below the tree line in winter and move onto the tundra in summer. The Kohistani people of Pakistan, for example, move from 2,000 feet (600 meters) in winter to 14,000 feet (4,300 meters) in summer. Swiss farmers move their goats to the tundra to graze during the summer. While there, the farmers raise some crops and make butter and cheese.

The native lifestyles discussed in this chapter represent traditional ways of life. Most tundra peoples, however, have adopted more modern lifestyles.

IMPACT OF THE TUNDRA ON HUMAN LIFE

The Arctic tundra is a harsh land, and life is shaped by the bitter cold. For humans, a few minutes' exposure to the cold can lead to frostbite and the possible loss of fingers, toes, noses, or ears. A longer exposure can lead to hypothermia (a lowering of the body temperature) and death.

Because people can change very little physically to adapt to tundra conditions, they must change their behavior. This is done by living a lifestyle in which nature is respected, people are creative, and nothing is wasted. For example, when the Inuit peoples of North America kill a caribou for food, they also use the skin for tents, bedding, and clothes. Harpoons are fashioned from the antlers, and needles and other tools are made from the bones. Thread for sewing is made from the animal's tendons.

Food Unlike people in more temperate and fertile biomes where the growing season is much longer, the people of the Arctic tundra do not rely on growing grains for food. They find their food by hunting and gathering. Some peoples, such as the Sami, who live in Scandinavia and Russia, survive by herding animals. Their diet consists primarily of meat and fish, and they hunt caribou and freshwater fish such as char. They also turn to the ocean for food, hunting whales, walruses, and seals. Berries and herbs supplement the diet, and leaves are used for medicine. The traditional diet may also be supplemented by canned goods shipped in during the summer months.

Alpine tundra dwellers hunt wild mountain animals for food, such as sheep, goats, and red deer. They also use llamas and yaks, domesticated herd animals, as food sources. South American Indians raise llamas for milk. Because the alpine climate is more mild, crops such as barley and potatoes are grown.

Shelter Traditionally, people of the tundra have built their homes from the materials at hand—frozen blocks of snow, pieces of sod, animal skins, or

A Sami farmer with his reindeer herd. At one time, reindeer herding was one of the traditional jobs for people living on the tundra. (Reproduced by permission of Corbis. Photograph by Farrell Grehan.)

stone. The Sherpas of the Himalayas build their houses of stone, making very thick walls. They live on top of the yak stables so that the warmth given off by the animals helps warm the home.

Clothing Tundra dwellers wear layers of clothing during the cold weather. If traditional, these clothes are made from the skins of animals. Caribou skin is a favorite choice because it is very warm yet light. During winter, two pairs of pants are worn. The inside pair is made with the animal hair facing in to keep the wearer warm. The outside pair is made with the hair facing outward to provide waterproofing.

People on the tundra may also wear parkas, which are jackets with hoods. The lining of the hood is usually wolf or wolverine hair. Sealskin is desirable for boots since it is waterproof. Bird skin lines the boots to make them warm. Moss, which is very absorbent, may be dried out and used for baby diapers.

Economic factors For traditional hunter-gatherers, possessions are almost meaningless. They do not have traditional jobs or seek to gain wealth. They spend their time finding the food they need and making clothing, weapons, and tools. In the past, reindeer herders were totally self-sufficient, getting food, clothing, and shelter from their herds. After World War II (1939–45), however, the economy of many tundra regions changed.

Many minerals have been discovered in the Arctic tundra. Coal, iron, nickel, gold, tin, and aluminum are found in eastern European countries such as Russia. Lead and zinc are mined in Greenland and gold in Canada. The largest mining activity on the tundra is drilling for natural gas and oil. Oil found on the tundra changed the economy of Western nations, who no longer need to rely so heavily on the Middle East for their oil supplies. The discovery of oil and the need to build a pipeline to move it brought many non-native people to the Alaskan tundra. The creation of jobs was also a boost to the economy of the countries that provided the work force.

After World War II, the Arctic tundra was also the site of military activity. In North America, the United States built a series of radar stations from Alaska to Greenland called the Distant Early Warning (DEW) Line. This brought new residents and new jobs to the area.

IMPACT OF HUMAN LIFE ON THE TUNDRA

Because of its relative isolation, either in the far north or high in the mountains, the tundra biome has not been affected as much by the presence of humans as have many other biomes. Its harsh environment has also helped keep people away. This is changing, however.

Use of plants and animals When native peoples use plants and animals only for their own food and necessary materials, wildlife populations remain stable. When people hunt with guns for commercial reasons, however, more animals are killed than are born. In the 1800s and throughout much of the 1900s, tundra animals were overhunted. For example, musk oxen were almost wiped out when their meat was sold to sailors on whaling ships. The skins were traded and the baby musk oxen sold to zoos. Caribou were also so severely overhunted that their herds were reduced by 90 percent.

Many countries have now set limits on the number of tundra animals that can be killed each year, including caribou, musk oxen, and polar bears.

Quality of the environment The quality of the Arctic tundra environment has been threatened by the effects of mining, the use of pesticides, and air pollution.

EFFECTS OF MINING Mining and drilling operations pollute the air, lakes, and rivers, as well as damage the land. In Russia, for example, the land around some nickel mines has become so polluted that all the plants have died. With no plant life to anchor it, the soil has washed away. Delicate plants are trampled as roads, airstrips, and houses are constructed. These plants take years to grow back, and some never recover. When mines are abandoned, all the old equipment and debris may be left behind to become eyesores.

> ## THE WHEELS THAT DUG LAKES
> The tundra environment is very fragile and its ability to restore itself is limited. When a construction truck or bulldozer is driven on the tundra during the summer it leaves ruts in the ground. When the Sun hits these ruts it causes the permafrost to melt, which then causes erosion, and the ruts get bigger. Each summer more of the exposed permafrost melts and eventually the rut turns into a gully. During World War II (1939–45), heavy trucks were driven on the tundra, leaving large ruts behind. The areas of permafrost that melted because of these ruts have grown so large over the years that some of them are now lakes.

Animals will not go near mining and drilling operations because of the noise and activity. This may leave them cut off from familiar food and water sources. The Alaskan pipeline, for example, was built right across the caribou migration route. In order to correct this problem, the pipeline has been raised like an overpass so they can pass underneath, but the road beside the pipeline may also be a problem. Scientists do not know what long-range effects these man-made additions will have on the animals.

The increasing number of people moving to the tundra to work in mining operations has created a need for houses, towns, and additional roads to bring in supplies. Getting rid of waste is also a serious problem. Because of the cold weather, garbage will not decompose and cannot be buried in the frozen ground. The choices are shipping it out, which is very expensive, or creating above-ground garbage dumps, which destroy the beauty of the landscape.

USE OF PESTICIDES Pesticides (poisons) have also affected tundra wildlife. A migratory bird such as a goose may eat plants sprayed with pesticides in the United States, where it spends the winter. When it returns to the Arctic in summer, it may be killed by a predator such as a fox. The fox and her babies eat the goose. A falcon or polar bear might eat one of the baby foxes. In this way, the pesticide contaminates a number of animals.

Peregrine falcons and polar bears have been especially affected by pesticide use. Pesticides make falcon eggs soft, and they break open before the chick has a chance to mature. Pesticides that have built up in the bodies of polar bears after they have eaten a number of contaminated animals, can ultimately kill them.

AIR POLLUTION The quality of the environment is also affected by what happens outside the tundra. Air pollution may travel to the tundra from thousands of miles away. In 1986, a nuclear reactor blew up at the Chernobyl power station in Prypíyat, near Kiev in the Ukraine. Nuclear waste was carried on air currents all over Europe and landed on plants in the tundra. When it reached Lapland tundra, reindeer ate the contaminated plants, got sick, and had to be destroyed.

Air pollution has also affected the ozone layer. Ozone is a chemical that forms naturally high up in the Earth's atmosphere. The ozone helps to protect 7the Earth from the harmful ultraviolet rays in sunlight. Pollution destroys the ozone, without which people, animals, and plants can suffer severe burns and develop cancers. The density of the ozone layer over the Arctic has been changing since the 1970s and is now greatly reduced.

The change in the ozone layer may also be affecting the world's climate. Carbon dioxide and other gases built up in the Earth's atmosphere traps heat from escaping. Called global warming, this may cause a change in temperature, melt glaciers, and cause the level of the sea to rise. The permafrost may also begin thawing, which would cause flooding and erosion, ultimately changing the landscape. Animal and plant life would be affected as habitats change.

> ## TRAVEL ON THE TUNDRA
>
> People who live on the tundra today use cars, airplanes, and snowmobiles. But methods of travel were not always so modern. Hunter-gatherers walked. Herders walked or rode their animals. The Laplanders rode in carts or sleighs pulled by reindeer. The Inuit invented a type of sled that is pulled by dogs, usually huskies. In summer, when the ice melted, the Inuit traveled in boats called kayaks, which could be made from sealskin spread over animal bones or wood.

NATIVE PEOPLES

Alpine tundra is home to the South American Indians who live in the Andes Mountains in Columbia, Ecuador, Peru, Bolivia, Chile, and Argentina. The Himalayan Mountains in Nepal are home to the Sherpas.

Arctic tundra dwellers include the Aleuts, Lapps, and Sami of Scandinavia and Russia, the Chukchi of eastern Siberia, and the Inuit of North America.

Traditionally, the Inuit of the Arctic tundra build snow houses (called igloos) when they need a temporary home while hunting. Their more permanent homes are built of bones and stones, with moss for roofing. The bladder of a seal or walrus is stretched across a hole in the wall to serve as a window. Platforms for sleeping are built up off the ground and lined with seal skin. Homemade lamps made of soapstone, which burn seal and whale blubber, give off both heat and light.

Traditionally hunters of seals, whales, and caribou, the Inuit people have been affected by the presence of military stations in the north, where many got jobs building and later rebuilding these stations.

THE FOOD WEB

The transfer of energy from organism to organism forms a food chain. All the possible feeding relationships that exist in a biome make up its food web. In the tundra, as elsewhere, the food web consists of producers, consumers, and decomposers. An analysis of the food web shows how energy is transferred within the tundra.

Green plants are the primary producers in the tundra. They produce organic materials from inorganic chemicals and outside sources of energy, like the Sun. Tundra annuals and hardy perennials, such as buttercups and dwarf willows, turn the Sun's energy into plant matter through photosynthesis.

An Inuit family working on their igloo. Traditionally people who lived on the tundra built their homes out of frozen blocks of snow as temporary homes while hunting. (Reproduced by permission of UPI/Corbis-Bettmann.)

Animals are consumers. Plant-eating animals, such as certain insects, caribou, reindeer, mountain goats, pikas, marmots, waterfowl, and lemmings, are primary consumers in the tundra food web. These animals then become food for the secondary consumers, which include predators such as spiders, wolves, and foxes. Tertiary consumers, such as grizzly bears and humans, eat other predators. Some tertiary consumers are omnivores.

Decomposer organisms, like bacteria and fungi, eat the decaying matter from dead plants and animals and release their nutrients back into the environment. Decaying matter in the soils provides nutrition for flowering plants and grasses, as well as for the fungi and bacteria themselves.

SPOTLIGHT ON TUNDRAS

NIWOT RIDGE

Niwot Ridge, an area of alpine tundra, is located about 22 miles (35 kilometers) west of Boulder, Colorado, and more than 9,843 feet (3,000 meters) above sea level. The ridge and the adjoining Green Lakes Valley cover approximately 3.8 square miles (10 square kilometers.) Runoff from the mountains feeds the Colorado and Mississippi Rivers.

Alpine tundra in the Rocky Mountains generally begins forming from 4,000 to 10,000 feet (1,219 to 3,048 meters) above sea level. Since its elevation is fairly high, Niwot Ridge is characterized by low temperatures throughout the year. The annual mean temperature is 5°F (-3.7°C). Temperatures in January average about 8°F (-13.2°C) and in July about 47°F (8.2°C).

This tundra is surrounded by subalpine forest at the lower elevations. Where the forest and tundra meet, subalpine meadows and patches of krummholz (dwarf trees) exist. Other physical features include a cirque glacier, a type of U-shaped glacier with its open end facing down the valley. Talus slopes formed by the accumulation of rock fragments are also found here.

> **NIWOT RIDGE**
>
> **Location:** Rocky Mountains west of Boulder, Colorado
>
> **Area:** 3.8 square miles (10 square kilometers)
>
> **Classification:** Alpine tundra

Some areas are snow covered. Strong winds that occasionally reach 170 miles (273 kilometers) per hour blow the snow from other areas, leaving them bare. Meltwater (water melted from ice or snow) is found only where there is snow. Although precipitation in summer is uneven, a beautiful array of yellow, pink, and purple flowers color the tundra in warm months. Plants include snow buttercup, old-man-of-the-mountain, Parry's primrose, and shooting stars.

Many birds live on this tundra in summer only. These include the water pipit, horned lark, and white-crowned sparrow. The only year-round residents are the white-tailed ptarmigan and the rosy finch.

Thirty-two species of mammals live on Niwot Ridge. Small plant-eating mammals include deer mice, voles, golden-mantled ground squirrels, pikas, yellow-bellied marmots, snowshoe hares, and porcupines. Badgers and weasels also are seen from time to time.

Many scientists and students work in the area, studying the weather, analyzing the soil, and observing plant and animal life. Niwot Ridge is designated as a Biosphere Reserve by the United Nations' Educational, Scientific, and Cultural Organization (UNESCO) and an Experimental Ecology Reserve by the United States Department of Agriculture (USDA) Forest Service.

The ridge is also part of the Roosevelt National Forest. When visitors are allowed on the tundra, they are encouraged to use the trails designated for this purpose because even one footprint can damage the fragile landscape.

BERING TUNDRA

The Bering Tundra is a western extension of the arctic coastal plain, a broad lowland in western Alaska between Kotzebue and Norton Sound. It is located on the Seward Peninsula on the American side of the former Bering Land Bridge (a narrow strip of land with water on both sides that at one time connected two continents) and is usually in the form of snow.

The Bering Tundra has cold winters and cool summers. Although summer temperatures in a polar climate do not usually exceed 50°F (10°C), a high of 90°F (32°C) in summer has been recorded here. In winter, the low has reached -70°F (-57°C). Annual precipitation averages 17 inches (43 centimeters).

Thousands of shallow lakes and marshes (wetlands with poorly drained soil and nonwoody plants) are found along the coast. Two large rivers, the lower Yukon and the Kuskokwim, flow out of the province to the southwest. The terrain on the peninsula varies from lava fields to hot springs to tundra.

> **BERING TUNDRA**
>
> **Location:** Seward Peninsula, Alaska
>
> **Area:** 23,400 square miles (60,610 square kilometers)
>
> **Classification:** Arctic tundra

Much of the tundra is less than 1,000 feet (305 meters) above sea level. Some small mountain groups range from 2,500 to 3,500 feet (762 to 1,067 meters) high. The highest point is Mount Osborn, which reaches a height of 4,714 feet (1,436 meters).

Permafrost lies under most of the area and the active layer of soil is considered young and undeveloped. Despite the permafrost and a growing season that can be as short as two weeks, there is still a wide variety of vegetation.

Species of dwarf trees, like birch, border the tundra. Birch, willow, and alder thickets are found between shoreline and forest. The lower Yukon and Kuskokwim Valleys are dominated by white spruce and cottonwood. Typical tundra plants found here include sedges, lichens, mosses, and cottongrass tussocks. Labrador tea, cinquefoil, and brightly colored forbs also grow in the region.

The coast provides habitats for migrating waterfowl and shore birds. Other bird species include ospreys, falcons, grouse, ravens, golden eagles, and various hawks and owls.

Mammals living in the tundra include musk oxen, brown and black bears, wolves, wolverines, coyotes, and a large herd of moose. Snowshoe hares, red foxes, lynxes, beavers, and squirrels summer here. Polar bears, walruses, and arctic foxes are sometimes seen along the northern coast of the Bering Sea.

Caribou migrate every summer to the tundra to give birth to their calves. They sometimes mix with domestic caribou, causing the native Eskimo herders to lose many of their animals, which join the wild herds.

An Inupiat Eskimo settlement of about 170 people is located in Wales on the western tip of the Seward Peninsula. Fewer than 60 miles (96 kilometers) from Siberia, Wales is one of the Alaskan settlements closest to Russia. Here, and in other places on the peninsula, native hunter-gatherers still live off the land. However, they are now faced with the challenge of adapting to changes caused by development, especially by the exploitation of natural resources such as oil.

During the early twentieth century, thousands of non-native people came to Alaska in search of gold. As a result, more than a million dollars in gold was taken from the peninsula.

KOLA PENINSULA

Murmansk is a province in northwestern Russia on the Kola Peninsula, a peak of high land that juts out into the ocean between the Barents and White Seas. Most of the peninsula lies across the Arctic Circle. It extends for about 190 miles (305 kilometers) from north to south, and 250 miles (400 kilometers) from east to west. Elevations in the peninsula's Khibiny Mountains reach 3,907 feet (1,191 meters), and Arctic tundra covers the northern areas.

As with many Arctic climates, winters are extremely severe and summers are short and cool. Summer temperatures in a polar climate do not usually exceed 50°F (10°C). Winter temperatures can reach -40°F (-40°C) or lower.

Many minerals are found on the peninsula, including the world's largest deposits of apatite, a mineral that is rich in phosphorus and used for fertilizer production. Nephelinite (a source of aluminum), zirconium, iron, and nickel are also mined here.

Bogs (wetlands with wet, spongy, acidic soil) are widespread and form in areas where the soil is saturated by water. Since summer meltwater cannot drain into the permafrost and the temperatures are too cool for evaporation, conditions are just right for the formation of bogs.

Despite problems with permafrost and a top-soil that is thin and poorly developed, the peninsula has been called a "botanical garden." Mosses, lichens, and dwarf Arctic birch cover most of the region. A forested area in the south has birch, spruce, and pine. Other typical tundra vegetation includes lichens, Lapp rhododendrons, Arctic willows, and white mountain avens, a yellow-flowered member of the rose family.

> ## KOLA PENINSULA
> **Location:** Murmansk Oblast Province, northern Russia
> **Area:** 40,000 square miles (100,000 square kilometers)
> **Classification:** Arctic tundra

Bird life typical to the tundra includes Siberian jays, Siberian titmice, grouse, and ptarmigans. Migrating seabirds include eider ducks, skuas, gulls, and Atlantic puffins. Lemmings, beavers, otters, and brown bears are common mammals, and thousands of reindeer migrate to the tundra in summer.

The largest town is the ice-free port of Murmansk, on the eastern shore, with a population of more than 1,000,000 people. Fishing is the main occupation along the coast. However, the most important part of the economy is mining.

In the interior, a few thousand Sami are engaged in reindeer herding, which was the basis of their economy until the twentieth century. Families lived in tents and migrated with their herds. This way of life is disappearing as families now have permanent homes and only the herders move with the reindeer.

The western part of Kandalaksha Bay and the area south of Murmansk are wildlife reserves. No visitors are allowed in these areas, which are headquarters for scientific research.

NORTHEASTERN SVALBARD NATURE RESERVE

The Svalbard Reserve is one of the largest and most important nature reserves in Norway. The area became protected by the Norwegian government in 1973 and includes North East Land, Kvit Island, Kong Karls Land, smaller adjoining islands, and the surrounding territorial waters. North East Land, the largest part of northeastern Svalbard, is covered by glaciers and ice caps all year long. "Svalbard" means "the land with the cold coast."

Arctic tundra covers the reserve, which is characterized by high winds, extremely cold temperatures, and permafrost, which reaches depths of 1,640 feet (500 meters). Only the upper 6 to 10 feet (1.8 to 3 meters) of ground thaw in the summer, so it is impossible for trees or plants with deep roots to grow here.

In most Arctic climates, winters are extremely severe, and summers are short and cool. However, a branch of the warm ocean current, called the North Atlantic Drift, moderates the climate here and temperatures range from 59°F (15°C) in the summer to -40°F (-40°C) in the winter.

The growing season is very short, lasting only a few weeks in the summer. Plant life is typical for Arctic tundra and includes more than 150 species of plants and ferns. There are many lichens and mosses, and tiny polar willows and dwarf birches grow along the tundra borders.

> ## NORTHEASTERN SVALBARD NATURE RESERVE
>
> **Location:** Svalbard Islands, Norway
>
> **Area:** 7,350 square miles (19,030 square kilometers)
>
> **Classification:** Arctic tundra

Almost 20 species of birds nest on the islands. These include murres, diving shorebirds with stocky bodies, short tails, wings, and webbed feet. Several types of gulls, ptarmigans, arctic terns, and phalaropes also nest here. Two species of ptarmigan, the rock ptarmigan and the willow ptarmigan, are the only birds that live here year round.

There are fewer species of mammals than birds in the area. Polar bears, Arctic foxes, reindeer, and musk oxen live in the reserve. Polar bears in this part of the world eat only meat, and seals are a big part of their diet. Not native to the reserve, the musk ox was imported from Greenland in 1929.

Since trapping is an important economic activity, land game, such as foxes, reindeer, and bears, is protected by law so it does not become endangered by overhunting. Hunters once lived on the tundra, but very few now spend the winter there. Most live in permanent communities at lower elevations. Svalbard's economy now depends upon industrial operations and coal mining.

Troms University, in co-operation with the University of Alaska, runs a Northern Lights research station in the reserve.

KATMAI NATIONAL PARK AND PRESERVE

The Katmai National Park and Preserve is located on the northeast coast of the Alaska Peninsula, along Shelikof Strait. The northern part of the park is Alpine tundra. Moving south, the terrain changes to forests, which line the bays and fiords (narrow inlets or arms of the sea bordered by steep cliffs).

The Katmai tundra begins at a relatively low elevation, around 2,000 to 2,300 feet (600 to 700 meters) above sea level. The highest peaks in the park reach 7,585 feet (2,312 meters).

The climate is cold and windy with harsh winters; short, cool, summers; and constant winds. In Katmai, the average summer temperature reaches only 60°F (15.5°C). In winter there is only about six hours of sunlight each day.

The terrain in this park is varied and includes glaciers, waterfalls, and mountains. The Valley of Ten Thousand Smokes, where once thousands of steam vents came through the valley floor, was formed in 1912 when the Novarupta Volcano erupted violently. Only a few active vents remain today, however.

Forests of blue spruces, alders, and willows border the tundra. Higher up, heaths, mountain avens (a member of the rose family), and bearberries grow.

Mammals living here include brown bears, wolves, foxes, moose, and caribou. The park is home to the largest population of protected brown bears in the world. The Alaskan brown bear, also known as the grizzly bear, is the world's largest carnivore and feeds on red salmon that spawn in the area.

> **KATMAI NATIONAL PARK AND PRESERVE**
>
> **Location:** Alaska
>
> **Area:** 5,806 square miles (15,038 square kilometers)
>
> **Classification:** Alpine tundra

People do not live here permanently because it is too inhospitable, but many visitors come to camp.

TAYMYR

Taymyr (also spelled Taimyr or Tajmyr) is a province located in northeastern central Russia. Mostly Arctic tundra, it extends from the Taymyr Peninsula south to the northern edge of the Central Siberian Plateau. The area includes the Severnaya Zemlya archipelago (a group of many islands) in the high Arctic.

The climate in Taymyr is exceptionally severe, with prolonged, bitter winters. The average temperature in January is -22°F (-30°C) and in July from 35° to 55°F (2° to 13°C.) There are few sunny days. Precipitation ranges from 4 to 14 inches (11 to 35 centimeters) annually.

> **TAYMYR**
>
> **Location:** North Central Russia
>
> **Area:** 332,850 square miles (862,100 square kilometers)
>
> **Classification:** Arctic tundra

The region has a variety of landforms, including mountains, lowlands, and plateaus (raised, flat-surfaced areas). The mountains of Byrrang are at an elevation of 3,760 feet (1,146 meters). In the Arctic, plateaus of ice can be as thick as 4,900 feet (1,500 meters). This ice was formed when glaciers covered much of the Arctic, more than 10,000 years ago. The Plateau of Plutoran has an elevation of 4,399 feet (1,341 meters).

A variety of minerals are found in Taymyr, including complex ores such as copper, nickel, platinum, gold, and coal. Gas is also mined.

The soil in Taymyr is typical tundra soil. The active layer of topsoil above the permafrost thaws in summer and freezes in winter. Bogs are widespread and form in areas where the soil is saturated with water. Since summer meltwater cannot drain into the permafrost and the temperatures are too cool for evaporation, conditions are just right for bog formation.

Even though the growing season is only a few weeks long, plants take advantage of the long hours of daylight during the short summer months. Most of the area is covered with mosses, sedges, rushes, and some grasses. Lichens grow on the hillsides and bilberries, a type of blueberry, grow in clusters. Many flowering plants provide color. On the edge of the tundra, dwarf willows and birch trees grow.

Taymyr is the summer home for the red-breasted goose, which is a threatened species. Steller's eider, a large sea duck, is also found on the peninsula. The mammal population includes reindeer, sable, wolf, fox, and white hare. Farther north live polar bears.

Few people live above the Arctic circle. Those that do include Russians, Dolgans, Nenets, Ukrainians, and Nganasans. Their chief economic activities are hunting, reindeer herding, fishing, and fur farming. Animals whose fur is used for clothing, such as the sable, are raised domestically rather than trapped in the wild. Only two cities are located in Taymyr province, the capital, Dudinka, and the port of Dikson. Their combined population is estimated at about 55,000.

Many of the traditional native lifestyles are slowly dying out. The Dolgan, for example, who used to be primarily reindeer herders, lived a nomadic lifestyle for the last several hundred years. However, since the 1970s, their lifestyle has become less nomadic. Now they rely on gardening as well as hunting for food.

The Russian government has established the Great Arctic Reserve on the Arctic tundra for shorebird conservation and research. The area is the breeding zone for many shorebirds who summer in Africa or along the Atlantic coasts of Europe.

GASPÉ

The Gaspé (Gas-PAY) Peninsula lies in eastern Quebec province, Canada. The forests and alpine tundra extend east-northeastward for 150 miles (240 kilometers) from the Matapédia River into the Gulf of St. Lawrence. The name Gaspé is probably from the Mi'kmaq Indian word "gespeg," which means "end of the world."

The tundra lies in the Chic-Choc, or Shickshock, Mountains, which are part of the Appalachian range. They are the highest in Quebec, with Mount Jacques Cartier rising to 4,160 feet (1,268 meters). The tundra on the Gaspé Peninsula begins around 3,280 feet (1,000 meters) above sea level.

Gaspé lies between a humid continental climate and a subarctic climate. Continental climates are warmed somewhat in summer as tropical air moves north. In winter, however, the cold polar air moves south. Gaspé faces severe winters, which have kept the population sparse. Temperatures can range from lows of -11° to 14°F (-24° to -10°C) in January to highs of 52° to 68°F (11° to 20°C) in July.

The land is crisscrossed by a number of rivers, and the peninsula is surrounded on three sides by the St. Lawrence River and the Gulf of St. Lawrence. A coniferous (evergreen) separates the province's deciduous (trees that loose their leaves at the end of each growing season) forests to the south from the subarctic and tundra areas, where no trees grow.

> **GASPÉ**
>
> **Location:** Eastern Quebec, Canada
>
> **Area:** 11,390 square miles (29,500 square kilometers)
>
> **Classification:** Alpine tundra

Arctic-alpine plants, which grow along the cliffs of the peninsula, include lichens, mosses, sedges, grasses, and other low woody and leafy plants. On the edges of the tundra maples, yellow birches, white spruces, and balsam firs grow.

Along the cliffs many seabirds can be found, including black-legged kittiwakes, double-crested cormorants, black guillemots, and razorbills. This is the only area of Quebec where caribou, moose, and white-tailed deer are found. Other mammals include foxes, lynxes, black bears, beavers, Arctic hares, and porcupines.

People do not live in this alpine tundra. It is a place for visitors and research only. The rest of the peninsula is sparsely populated with about one-fifth of the people earning a living through agriculture. In summer months, many tourists visit the peninsula and support the economy. Lumbering, mining, and fishing are the major economic activities. Minerals mined in the area are copper, lead, and zinc.

The Mi'kmaq (also spelled Micmac) people, who have inhabited the Gaspé Peninsula for hundreds of years, were one of the eight major tribes who comprised the Woodland First Nations peoples. Traditionally, they were a seasonally nomadic people, hunting moose, caribou, and small game in winter and fishing in summer. Their winter homes were conical (cone-shape) wigwams covered with birch bark or animal skins. In summer, they lived in open-air, oblong wigwams. Homes were portable and easy to put up or move. Travel was by canoe, toboggan, or snowshoe. Different types of snowshoes were used, depending on the type of snow. Some of the Mi'kmaq still live on Gaspé, but many have migrated to the United States.

The Gaspésian Provincial Park is a large conservation area covering much of the peninsula. Another park, Forillon National Park, covers 93 square miles (240 square kilometers) at the northeastern tip of the peninsula.

ARCTIC NATIONAL WILDLIFE REFUGE

The Arctic National Wildlife Refuge in Alaska is one of the largest wildlife refuges in the world. To the north is the Arctic Ocean and to the south is the Porcupine River. The refuge extends east to west more than 200 miles (322 kilometers) from the Trans-Alaska pipeline corridor to Canada, and almost 200 miles north to south from the Beaufort Sea to the Yukon Flats National Wildlife Refuge. It supports both Arctic and alpine tundra.

> **ARCTIC NATIONAL WILDLIFE REFUGE**
>
> **Location:** Alaska
>
> **Area:** 19,000,000 acres (7,600,000 hectares)
>
> **Classification:** Arctic and alpine tundra

The Brooks Range, with peaks as high as 9,000 feet (2,743 meters), extends from east to west through the refuge. The Brooks Range is the highest mountain range within the Arctic Circle and is the northernmost extension of the Rocky Mountains in northern Alaska.

Winter on the refuge is long, and snow usually covers the ground at least nine months of the year. The average winter temperature here is below 0°F (below -17°C). The wind chill factor makes the temperature feel like -100°F (-73°C) to exposed skin. In summer, the temperature averages 50°F (10°C). Annual precipitation ranges from 4 to 16 inches (10 to 40 centimeters).

Besides two tundra zones, the refuge also includes barren mountains, forests, shrub thickets, and wetlands. The alpine tundra is crisscrossed by braided rivers and streams, with clusters of shallow, freshwater lakes and marshes.

The Arctic National Wildlife Refuge contains the greatest variety of plant and animal life of any conservation area within the Arctic Circle. Even though the growing season can be as short as two weeks, the continuous summer daylight helps plants grow rapidly. Reindeer moss, sedges, and flowering plants are common. Spruces, poplars, birches, and willow trees grow in areas surrounding the tundra.

More than 180 species of birds have been observed here. Peregrine falcons, endangered or threatened in the lower 48 states, are common in the refuge as are rock ptarmigans, grebes, snow geese, plovers, and sandpipers.

Thirty-six species of land mammals live here, including all three species of North American bears (black, brown, and polar). Smaller mammals include lynxes, wolverines, lemmings, Arctic hares, and wolves.

In the summer, tens of thousands of caribou give birth to and raise their calves on the tundra. These caribou migrate south in winter to northeastern Alaska and the northern Yukon, almost 1,000 miles (1,609 kilometers), farther than any other land animal. Even though their winter habitat is more mild than their tundra breeding grounds, they still face Arctic weather. The caribou can dig through snow almost 2 feet (60 centimeters) deep to reach food, which consists of lichens, sedges, and grasses. The refuge protects most of the calving grounds for the Porcupine caribou herd, the second largest herd in Alaska, which numbers about 180,000 animals. It also protects a large part of their migration routes.

Dall sheep live in the refuge all year, mostly in the tundra, staying close to rocky outcrops and cliffs where they are safe from predators which include wolves, golden eagles, bears, and humans. Dall sheep eat grasses, sedges, broad-leaved plants, and dwarf willows. In winter, when these foods are scarce, they eat lichens.

The area is home to several groups of native peoples and is the site of controversy over oil development. Kaktovik, an Inupiat Eskimo village, borders the refuge on the north side. The Inupiat have incorporated and, along with the government, own part of Alaska. The Gwich'in, another native group, are one of the few who have not incorporated and choose to maintain their reservation status and a subsistence lifestyle in their nine villages.

The Inupiat see the value of oil development in the area. It provides jobs and supports schools and community improvements. The Gwich'in see it as endangering the fragile tundra environment and disturbing migration routes and breeding places for the caribou.

HUASCARAN NATIONAL PARK

Huascaran (hwa-SCAH-run) National Park is a place of biological diversity. Situated in the Andes Mountains in Peru, the landscape is dominated by mountain peaks, glaciers, lakes, and alpine tundra. It became a national park in 1975 and a World Heritage Site in 1985. The Quechua (QWAY-chwah) people, who live near the park, maintain a traditional lifestyle.

> **HUASCARAN NATIONAL PARK**
> **Location:** Peru
> **Area:** 840,000 acres (340,000 hectares)
> **Classification:** Alpine tundra

Alpine tundra, characterized by cold winds, is found on mountains above the tree line. The lower portion of the mountain, is covered by tropical rain forest.

The snowcapped peak of Mount Huascaran, the highest mountain in Peru, rises to an elevation of 22,205 feet (6,768 meters) above sea level. The climate varies, with temperatures getting colder and rainfall decreasing as the elevation increases. Generally, the tundra climate has an average annual temperature below 50°F (10°C).

Much of Huascaran National Park is covered with permanent ice and snow. A permanent snowline begins between 14,700 and 16,400 feet (4,480 and 5,000 meters). Just below the tundra is the *puna,* an area of alpine grasslands with no trees. Below that is the *tierra fria,* the cool land, where most of the people live. This area has forest and land suitable for growing wheat and potatoes.

Vegetation is typical of alpine tundra in other parts of the world. Mosses and lichens grow on bare rocks. Short flowering plants and grasses, cushion plants, and sedges are also common.

The South American, or Andean, condor nests on the cliffs in the high tundra. A member of the vulture family, it is one of the largest flying birds in the world with a wing span of about 10 feet (3 meters). The condor is carnivorous and feeds on dead animals.

Mammals in the Andes are different from those in other alpine tundra. The llama, alpaca, vicuña, and guanaco are native to this part of the world. Although they are all relatives of the camel, they have no hump. The vicuña is an endangered species hunted for its fleece, from which some of the world's finest wool is made. The chinchilla, a small rodent also common here, was hunted almost to extinction for its fur.

South American Indians living near the Huascaran Park are the Quechua and the Aymara, who have adapted to cold, high-altitude living. They have larger lungs and hearts, and their blood contains more red blood cells, allowing it to hold more oxygen. This makes it easier for them to live and work in the thin atmosphere of the mountains. They are also able to walk barefoot on icy rocks. These native peoples lead isolated lives as herders and farmers. Women make clothes and blankets to sell at market, spinning their own wool and weaving fabrics.

FOR MORE INFORMATION

BOOKS

Fowler, Allan. *Arctic Tundra.* Danbury, CT: Children's Press, 1997.

Inseth, Zachary. *The Tundra.* Chanhassen, MN: The Child's World, Inc., 1998.

Pipes, Rose. *Tundra and Cold Deserts.* Orlando, FL: Raintree Steck-Vaughn Publishers, 1999.

Wadsworth, Ginger. *Tundra Discoveries.* Walertown, MA: Charlesbridge Publishing, Inc., 1999.

ORGANIZATIONS

Center for Environmental Education
  Center for Marine Conservation

1725 De Sales St. NW, Suite 500
Washington, DC 20036

Environmental Defense Fund
257 Park Ave. South
New York, NY 10010
Phone: 800-684-3322; Fax: 212-505-2375
Internet: http://www.edf.org

Environmental Network
4618 Henry Street
Pittsburgh, PA 15213
Internet: http://www.envirolink.org

Environmental Protection Agency
401 M Street, SW
Washington, DC 20460
Phone: 202-260-2090
Internet: http://www.epa.gov

Friends of the Earth
1025 Vermont Ave. NW, Ste. 300
Washington, DC 20003
Phone: 202-783-7400; Fax: 202-783-0444

Greenpeace USA
1436 U Street NW
Washington, DC 20009
Phone: 202-462-1177; Fax: 202-462-4507
Internet: http://www.greenpeaceusa.org

Nature Conservancy
1815 N. Lynn Street
Arlington, VA 22209
Phone: 703-841-5300; Fax: 703-841-1283
Internet: http://www.tnc.org

Sierra Club
85 2nd Street, 2nd fl.
San Francisco, CA 94105
Phone: 415-977-5500; Fax: 415-977-5799
Internet: http://www.sierraclub.org

World Wildlife Fund
1250 24th Street NW
Washington, DC 20037
Phone: 202-293-4800; Fax: 202-229-9211
Internet: http://www.wwf.org

WEBSITES

Note: Website addresses are frequently subject to change.

Arctic Studies Center: http://www.nmnh.si.edu/arctic

Long Term Ecological Research Network: http://lternet.edu

Multimedia Animals Encyclopedia: http://www.mobot.org/MBGnet/vb/ency.htm

National Geographic: http://www.nationalgeographic.com

National Science Foundation: http://www.nsf.gov

Thurston High School: Biomes: http://ths.sps.lane.edu/biomes/index1.html

BIBLIOGRAPHY

Aldis, Rodney. *Ecology Watch: Polar Lands*. New York: Dillon Press, 1992.

"Arctic National Wildlife Refuge." U. S. Fish and Wildlife Services. http://www.r7.fws.gov/nwr/arctic/arctic.html

Catchpoh, Clive. *The Living World: Mountains*. New York: Dial Books for Young Readers, 1984.

Claybourne, Anna. *Mountain Wildlife*. Saffron Hill, London: Usborne Publishing, Ltd., 1994.

Forman, Michael. *Arctic Tundra*. New York: Children's Press, 1997.

Fornasari, Lorenzo and Renato Massa. *The Arctic*. Milwaukee, WI: Raintree-Steck Vaughn Publishers, 1989.

George, Jean Craighead. *One Day in the Alpine Tundra*. New York: Thomas Y. Crowel, 1984.

Grzimek, Bernhard, ed. *Grzimek's Animal Life Encyclopedia: Fishes II and Amphibians*. New York: Van Nostrand Reinhold Co., 1972.

Grzimek, Bernhard, ed. *Grzimek's Animal Life Encyclopedia: Reptiles*. New York: Van Nostrand Reinhold Co., 1972.

Hiscock, Bruce. *Tundra: The Arctic Land*. New York: Atheneum, 1986.

Kaplan, Elizabeth. *Biomes of the World: Tundra*. New York: Marshall Cavendish, 1996.

Khanduri, Kamini. *Polar Wildlife*. London: Usborne Publishing, Ltd., 1992.

"Kola Peninsula." Britannica Online. http://www.eb.com:180/cgi-bin/g?DocF=micro/326/3.html

Lye, Keith. *Our World: Mountains*. Morristown, NJ: Silver Burdett, 1987.

Multimedia Animals Encyclopedia. http://www.mobot.org/MBGnet/vb/ency.htm

National Park Service, Katmai National Park and Preserve. http://www.nps.gov/katm/

Rootes, David. *The Arctic*. Minneapolis: MN: Lerner, 1996.

Sage, Bryan. *The Arctic and Its Wildlife*. New York: Facts on File, 1986.

Sayre, April Pulley. *Tundra*. New York: Twenty-first Century Books, 1994.

Shepherd, Donna. *Tundra*. Danbury, CT: Franklin Watts, 1997.

Silver, Donald. *One Small Square: Arctic Tundra*. New York: W. H. Freeman and Co., 1994.

Siy, Alexandra. *Arctic National Wildlife Refuge*. New York: Dillon Press, 1991.

Slone, Lynn. *Ecozones: Arctic Tundra*. Viro Beach, FL: Rourke Enterprises, 1989.

Steele, Philip. *Geography Detective: Tundra*. Minneapolis, MN: Carolrhoda Books, Inc., 1996.

"Taymyr." Britannica Online. http://www.eb.com:180/cgi-Bin/g?DocF=micro/584/53.html

University of Wisconsin, Stevens Point. http://www.uwsp.edu/acaddept/geog/faculty/ritter/geog101/biomes_toc.html

WETLAND

Wetlands are areas that are covered or soaked by ground or surface water often enough and long enough to support special types of plants that have adapted for life under such conditions. Wetlands occur where the water table (the level of groundwater) is at or near the surface of the land, or where the land is covered by shallow surface water (usually no deeper than about 6 feet [1.8 meters]). They form the area between places that are always wet, such as ponds, and places that are always dry, like forests and grasslands.

Many wetlands are not constantly wet, experiencing what is called a wet/dry cycle. Some are wet for only part of the year, like those that are drenched during heavy seasonal rains. Some have no standing water but, because they are near the water table, their soil remains saturated (soaked with water). Others may dry out completely for long periods, sometimes even for years.

Wetlands are among the world's most productive and biodiverse environments (environments that support a large variety of life forms). Only rain forests and coral reefs are more biodiverse.

Wetlands are found all over the world, in every climate—from the frozen landscape of Alaska to the hot zones near the equator—except Antarctica. They make up about 6 percent of the world's landmass. In the United States, there are about 240,000,000 acres (96,000,000 hectares) of wetland, and some type can be found in every state. Alaska alone contains about 170,000,000 acres (68,000,000 hectares) of wetlands.

HOW WETLANDS DEVELOP

Wetlands have a life cycle that begins with their formation and may involve many changes over time.

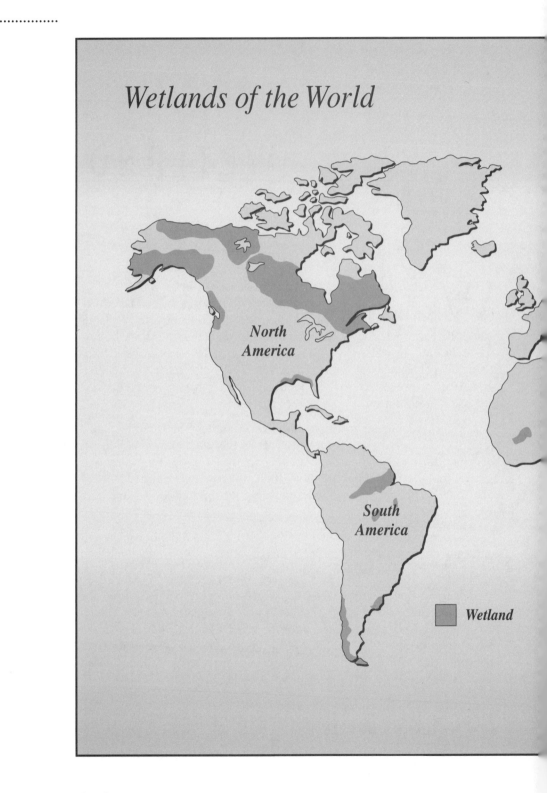

Wetlands of the World

North
America

South
America

Wetland

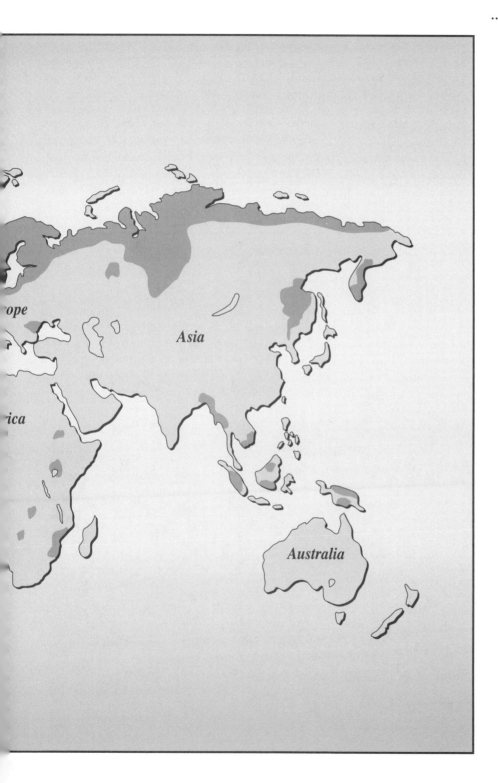

FORMATION

Many wetlands were formed when glaciers retreated after the last Ice Age, about 10,000 years ago. Some of the glaciers left depressions in the ground, called kettles, which were perfect places for water to gather. In some places, buried ice melted to form kettle lakes that eventually turned into wetlands.

Wetlands are formed by the overflowing of river banks and changes in sea level, which can leave waterlogged areas behind. (the average height of the sea) Some wetlands are formed with help from beavers making dams that cause rivers or streams to back up and flood the surrounding area. Landslides or centuries of heavy winds may also change the terrain or carve out depressions in the ground where water then collects.

Climate, too, plays a key role in the formation of wetlands. Lots of rain and little drainage, for example, can cause the ground to become waterlogged.

SUCCESSION

Wetlands are found all over the world, like this one in Somerset, England. (Reproduced by permission of Corbis. Photograph by Pat Jerrold.)

Wetlands are constantly evolving as their plant life changes. This process is called succession. For example, one type of floating plant, such as pondweed, begins to fill up a pond or lake. The waste from these plants, such as dead stems and leaves, makes the water thick, shallow, and slow moving. As a result, plants that must be anchored in soil, such as reeds and

grasses, can then grow. As waste matter continues to accumulate, the pond gradually becomes wetland. As succession continues, the wetland eventually disappears and is replaced by dry ground.

KINDS OF WETLANDS

The three main types of wetlands are swamps, marshes, and peatlands. They can be identified by size, the type of soil, and the plants that live in them, as well as by the type and amount of water they hold. Some located inland are fresh water, while coastal wetlands may be fresh or salt water. Many areas can contain several different types of wetlands. The Great Dismal Swamp in Virginia, which contains both peat bogs and freshwater swamps, is an example.

SWAMP

A swamp is a wetland that is characterized by poorly drained soil and plant life dominated by trees. Swamp soil is usually saturated for most of the

An illustration showing the stages in wetland succession.

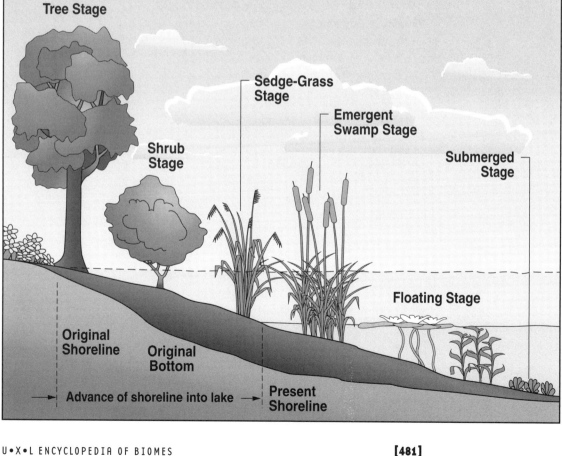

U•X•L ENCYCLOPEDIA OF BIOMES

year. The water can range from 1 inch (2.5 centimeters) to more than 1 foot (30 centimeters) deep.

Because swamp ground is constantly waterlogged, plant life is kept to a minimum. The trees and plants that grow there must be able to tolerate having their root systems wet for long periods of time. The water is stagnant (without movement) and the dead plant matter settles on the bottom and receives little oxygen. Without oxygen, the dead matter cannot fully decay. This gives the swamp its characteristic brown water and unpleasant smell.

Swamps are usually found in low-lying areas near rivers or along coastal areas. They get water from the overflow of the river or from ocean tides. Inland swamps always contain fresh water, but coastal swamps may contain either fresh or salt water.

Some swamps, called pocosins, form in upland areas. The term "pocosin" comes from the Algonquin phrase meaning "swamp on a hill." Rain is usually their only source of water, meaning the soil is low in minerals and nutrients which come from ground water. Drainage is poor in pocosins and the saturated soil is acidic (high in acids). In this environment, decomposition (breaking down) of waste is slow, and decaying matter accumulates over time. Sometimes pocosins are flooded by slow-moving streams.

Freshwater swamp Freshwater swamps develop near the edges of lakes and next to rivers that overflow their banks. Partly decayed plant matter and stagnant water soon create swamp conditions. The Everglades in Florida, the Okefenokee Swamp in Georgia, and the Great Dismal Swamp in North Carolina and Virginia, are examples.

Saltwater swamp A saltwater swamp is formed by the ebbing (receding) and flowing of the ocean tides. (The tides are a rhythmic rising and lowering of the oceans caused by the gravitational pull of the Sun and Moon.) When the tide is low, the flat areas of the swamp are not under water. Saltwater swamps can be found in New Guinea, Indonesia, Kenya, Zaire, along the Mekong River in Vietnam, and along the Ganges River in India.

In some low-lying coastal areas regularly flooded with sea water mangrove swamps develop. A mangrove swamp is a coastal, saltwater wetland that is found in tropical and subtropical climates such as southern Florida and Puerto Rico. (Tropical and subtropical climates are found in areas close to the equator and are characterized by warm weather.) The mangrove swamp supports salt-loving shrubs and trees such as the mangrove, from which it takes its name. Because of the high salt content, few other plants grow there.

MARSH

About 90 percent of all wetlands are marshes. Marshes are dominated by nonwoody plants such as grasses, reeds, rushes, sedges, and rice. Some marshes are so thickly covered with vegetation that they look like fields of grass.

Marshes are usually found in temperate climates, where summers are hot and winters are cold, and they often form at the mouths of rivers (the point where a river enters into a larger river or ocean), especially those having large deposits of soil and sand, called deltas. They also form in flat areas that are frequently covered by shallow fresh or salt water.

Freshwater marsh Freshwater marshes are the most common of the world's temperate wetlands. Some receive no rain, and most are found along the edges of lakes and rivers or where groundwater, streams, or springs cause flooding. The Camargue in the Rhone Delta in southern France is an example. The different types of freshwater marshes include prairie potholes, riparian areas, wet meadows, and washes.

Prairie potholes are small marshes that can be found by the millions throughout the north central area of the United States and south central Canada. Usually no more than a few feet deep, they may cover as much as 10,000 acres (4,000 hectares). Prairie potholes are important to migratory waterfowl as places to rest, feed, and nest. In Africa, these small marshes are called dambos. In North America it is estimated that 50 to 70 percent of all waterfowl are hatched in prairie potholes.

Wet meadows are freshwater marshes that frequently become dry. They look like grasslands but the soil is saturated. They are common in temperate and tropical regions around the world, including the midwestern and southeastern United States. Their dominant plant is the sedge, a flowering herb that resembles grass.

Riparian wetlands are marshes found along rivers and streams. They range in width from a few feet (meters) to as much as 12 miles (20 kilometers). Smaller riparian wetlands are common in the western United States. Larger ones are located along large rivers, such as the Amazon in South America. Riparian wetlands are unique from others in that the vegetation and life forms found immediately adjacent to the river or streams differ from those of the surrounding area, usually a forest. This marked contrast produces a diversity and enhances the benefits for wildlife, both in food (grasses and sedges) and in shelter (shrubs and trees) in a relatively small place.

A wash is a dry streambed that becomes a wetland only after a rain. Washes are found in dry plains and deserts. The plant life they support usually has a short growing season and disappears during dry periods. In North Africa and Saudi Arabia, washes are called *wadis*. In the American west they are called *arroyos*.

> ### THE LOST SOUL
>
> Sometimes at night strange lights can be seen flickering over marshes. Legend has it that the light is the soul of a person who has been rejected by Heaven and is condemned to carrying hot coals as he or she wanders around the Earth.
>
> This light goes by several names: Jack-o'-lantern, will-o'-the wisp, and foolish fire. The light is probably caused when marsh gas, produced by decaying plants, spontaneously catches fire.

Saltwater marsh Saltwater marshes, also called salt marshes, are found in low, flat, poorly drained coastal areas. They are often flooded by salt water or brackish water (a mixture of both fresh and salt water). Saltwater marshes are especially common in deltas, along low seacoasts, and in estuaries (arms or inlets of the sea where the salty tide washes in and meets the freshwater current of a river). They can be found in New Zealand, in the Arctic, and along the Atlantic, Pacific, Alaskan, and Gulf coasts.

Saltwater marshes are greatly affected by the tides, which raise and lower the water level on a daily basis. A saltwater marsh may have tidal creeks, tidal pools, and mud flats, each of which has its own ecosystem (a network of organisms that have adapted to a particular environment).

The high salt content of the sea water in the marsh makes it hard for plants to adapt. However, grasses such as marsh grass, cord grass, salt hay grass, and needlerush thrive here.

> ### THE CURSE OF ST. COLUMBA
>
> In Northern Ireland, some farmers still cut three steps into peat being dug from bogs for fuel. They do this to avoid the curse of St. Columba, who is said to have been trapped in a bog and who cursed all those who did not cut steps for his escape.

PEATLAND

Peatlands are wetlands in which peat has formed. Peat is a type of soil made up of the partially decayed remains of dead plants, such as sphagnum moss and even trees. In peatlands, dead plant matter is produced and deposited at a greater rate than it decomposes. Over time, sometimes over thousands of years, a layer of this plant matter is formed that may be as deep as 40 feet (12 meters). Several conditions are necessary for peat to form, however. The soil must be acidic, waterlogged from frequent rains, and low in oxygen and nutrients. Because the bacteria responsible for decomposing plant matter cannot thrive under these conditions, the layers of partially decayed plant matter accumulate.

> ### NATURE'S DIAPERS
>
> Sphagnum moss, which is found in bogs, can absorb many times its weight in water. At one time, certain Native Americans dried this moss and used it as diapers for their babies.

There are about 1,500,000 square miles (4,000,000 square kilometers) of peatland around the world. Peatlands are located in colder, northern climates as well as in the tropics and can be found in Russia, Scandinavia, northern Europe, England, Ireland, Canada, and the northern United States.

The types of peatlands are temperate bogs, fens, and tropical tree bogs.

Temperate bog Ninety percent of peatlands are found in northern temperate climates where they are called bogs. Bogs have a soft, spongy, acidic soil that retains moisture, which comes primarily from rainwater. Some bogs have taken as long as 9,000 years to develop and are sources of peat that can be burned as fuel.

Temperate bogs have been described as soft, floating carpets. The carpet is made mostly of sphagnum moss, which can be red, orange, brown, or green and may grow as long as 12 inches (30 centimeters). Sphagnum moss can grow in an acidic environment, and it holds fresh rainwater in which other bog plants can grow. Because of the high concentration of sphagnum moss, other bog plants have developed unusual adaptations to help them get nutrients. For example, the bog myrtle forms a partnership with the bacteria in its roots to obtain extra nitrogen. Besides the bog myrtle, other plants that grow in bogs include grasses, small shrubs such as leatherleaf, flowering plants such as heather, and poison sumac. Some rare wildflowers, such as the lady slipper orchid and the Venus' flytrap, are also found in bogs.

Most bogs lie in depressed areas of ground. Some, however, called raised bogs, grow upward and are an average of 10 feet (3 meters) higher than the surrounding area.

Blanket bogs, which are shallow and spread out like a blanket, form in areas with relatively high levels of annual rainfall. Their average depth is 8 feet (2.6 meters). Blanket bogs are found mainly on lowlands in the west of Ireland and in mountain areas. After a heavy rainfall, bogs that are located on steep slopes can wash down like a huge landslide of jelly and cover cattle, farms, and even villages.

A bright green cushion of curly sphagnum moss, common to bogs. (Reproduced by permission of Corbis. Photograph by Andrew Brown.)

Fen When a peatland lies at or below sea level and is fed by mineral-rich groundwater, it is called a fen. The characteristic plants in a fen are grasses, sedges, and reeds rather than sphagnum moss, and the soil does not become as acidic as in a bog. Therefore, less peat accumulates in a fen and it becomes only about 6 feet (2 meters) thick. As plant matter accumulates in a fen over time, it will form a raised bog.

Tropical tree bog Bogs found in tropical climates are called tree bogs. These bogs produce peat from decaying trees rather than from sphagnum moss. The tree most commonly found in these bogs is the broad-leaved evergreen. Temperatures are warmer than in temperate regions, causing decay to occur more rapidly. As a result, not as much peat develops. The only source of water is rain.

Tropical tree bogs can be found in South America, Malaysia, and Indonesia.

CLIMATE

Unlike some other biomes, wetlands do not have a characteristic climate. They exist in polar, temperate, and tropical zones, although usually not in deserts. However, they are very sensitive to changes in climate, such as a decrease in precipitation (rain, sleet, or snow). The amount of precipitation and changes in temperature affect the growth rate of wetland plants. Some wetlands are seasonal, which means that they are dry for one or more seasons of the year.

TEMPERATURE

Temperatures vary greatly depending on the location of the wetland. Many of the world's wetlands are in temperate zones (midway between the North and South Poles and the equator). In these zones, summers are warm and winters are cold, but temperatures are not extreme. However, wetlands found in the tropic zone, which is around the equator, are always warm. Temperatures in wetlands on the Arabian Peninsula, for example, can reach 122°F (50°C). In northeastern Siberia, which has a polar climate, wetland temperatures can be as cold as -60°F (-51°C).

RAINFALL

The amount of rainfall a wetland receives depends upon its location. Wetlands in Wales, Scotland, and western Ireland receive about 59 inches (150 centimeters) per year. Those in Southeast Asia, where heavy rains occur, can receive up to 200 inches (500 centimeters). In the northern areas of North America, wetlands exist where as little as 6 inches (15 centimeters) of rain fall each year.

GEOGRAPHY OF WETLANDS

The geography of wetlands involves landforms, elevation, and soil.

LANDFORMS

Landforms found in wetlands depend upon location, soil characteristics, weather, water chemistry, dominant plants, and human interference. Their physical features are often short-lived, especially if they are near floodplains or rivers, which can cause abrupt changes. Wetlands usually form in a basin where the ground is depressed, or along rivers and the edges of lakes.

ELEVATION

Wetlands can be found at many elevations (the height of an area in relation to sea level). Some wetlands in the Rocky Mountains in North America, for example, are at an elevation of 10,000 feet (3,048 meters).

Elevation is used to help classify some wetlands in Ireland. Bogs that are less than 656 feet (200 meters) above sea level are called Atlantic blanket bogs. Those that are more than 656 feet (200 meters) above sea level are called mountain blanket bogs.

SOIL

An important characteristic of a wetland is its soil. Soil composition helps to determine the type of wetland and what plants and animals can survive in it. Almost all wetland soils are at least periodically saturated.

Wetland soils are hydric. This means that they contain a lot of water but little oxygen. The plants that live in wetlands are only those that can adapt to these wet soils. The nutrients in the soil often depend upon the water supply. If the water source is primarily rain, the wetland soils do not receive as many minerals as those fed by groundwater. Soil in floodplains is very rich and full of nutrients, including potassium, magnesium, calcium, and phosphorus.

In some bogs associated with forests, decaying plant matter fully decomposes and is combined with sediments to form muck. This type of soil is dark and glue-like. To classify as muck, soil must contain not less than 20 percent organic (derived from living organisms) matter.

PLANT LIFE

One of the most important characteristics of any biome is its plant life or lack of plant life. More than 5,000 species of plants live in or near wetlands. Wetlands have high biological productivity (the rate at which life forms grow in a certain period of time). The higher its plant productivity, the more animal life a wetland can support. The kinds of plants that may be found in a wetland are determined by several factors, especially the type of soil and the quantity of water.

Some plants, called hydrophytes, grow only in water or extremely wet soil. Sedges are an example. Mesophytes, such as reeds, need moist but not saturated soil. When a wetland dries up, the area fills with plants called xerophytes. These are plants adapted to life in dry habitats and can survive where other wetland plants would wilt. They include water hemp, marsh purslane, and fog fruit.

A submergent plant grows beneath the water and is found in deep marshes and ponds. Even its leaves are below the surface. Submergents include milfoil, pondweed, and bladderwort, an insect-eating plant.

Found in deep marshes, floating aquatics float on the water's surface. Some, such as duckweed have free-floating roots. Others, such as water lilies, water lettuce, and water hyacinths have leaves that float on the surface, stems that are underwater, and roots that are anchored to the bottom.

An emergent plant grows partly in and partly out of the water. The roots are usually under water, but the stems and leaves are at least partially exposed to air. They have narrow, broad leaves, and some even produce flowers. Emergents include reeds, rushes, grasses, cattails, and water plantains.

A USEFUL NUISANCE

People often grumble about the water hyacinth because it multiplies and spreads quickly in open water. The plants become so thick that boats have a hard time moving through them. However, the water hyacinth is also useful. It absorbs and neutralizes many pollutants that would otherwise contaminate the water. It also has a high nutrient content, which makes it a good fertilizer.

ALGAE, FUNGI, AND LICHENS

It is generally recognized that algae (AL-jee), fungi (FUHN-jee), and lichens (LY-kens) do not fit neatly into the plant category. In this chapter, however, we will discuss these special organisms as if they, too, were plants.

Algae Most algae are one-celled organisms too small to be seen by the naked eye. They make their food by a process called photosynthesis. (Photosynthesis is the process by which plants use the energy from sunlight to change water and carbon dioxide from the air into the sugars and starches they require.) Some wetland algae drift on the surface of the water, forming a kind of scum. Others attach themselves to weeds or stones. Some can grow on the shells of turtles or inside plants or animals. Microscopic algae that can be found in saltwater marshes include diatoms and green flagellates (FLAJ-uh-lates). Desmids are a type of green algae found in bogs.

Fungi Fungi are plantlike organisms that cannot make their own food by means of photosynthesis; instead, they grow on decaying organic matter or live as parasites (organisms that depend upon another organism for food or other needs).

Fungi grow best in a damp environment, which makes wetlands a favorable home. Common wetland fungi include mushrooms, rusts, and puffballs.

Lichens Lichens are combinations of algae and fungi. The alga produces the food for both by means of photosynthesis, while it is believed the fungus absorbs moisture from the air and provides shade. One of the most common wetland lichens is called reindeer moss, an important food source for northern animals such as caribou.

GREEN PLANTS

Most green plants need several basic things to grow: light, air, water, warmth, and nutrients. In a wetland, light and water are in plentiful supply.

Nutrients—primarily nitrogen, phosphorus, and potassium—are obtained from the soil. Some wetland soils, however, are lower in these nutrients. They may also be low in oxygen. As a result, many wetland plants have special tissues with air pockets that help them to "breathe."

Trees that grow in swamps, such as the mangrove and bald cypress, have shallow roots. Because of the lack of oxygen in the soil, the roots remain near the surface. Other trees found in mild climates include willows and alders. Palms are found in warm climates.

Wetland plants are also classified as submergents, floating aquatics, or emergents, according to their relationship with water.

Growing season Climate and precipitation affect the length of the growing season in a wetland. Warmer temperatures and moisture usually signify the beginning of growth. In regions that are colder or receive little rainfall, the growing season is short. Growing conditions are also affected by the amount of moisture in the soil, which ranges from saturated to dry. Although many wetland plants die during dry periods, their seeds remain in the soil all winter and sprout in the spring.

> ### THE HAT THROWER
> An unusual fungus called a hat thrower grows on dung (animal waste) deposited in wetlands. The fungus forms a tiny black bulb that looks like a hat. The hat explodes under pressure and pieces get thrown outward from the fungus and attach to other plants. When an animal eats those plants, the hat thrower passes through its digestive system. When the animal defecates, a new fungus grows on the pile of dung.

Reproduction Green wetland plants reproduce by several methods. One is pollination, in which the pollen from a plant's male reproductive part, called the stamen, is carried by wind or insects to the plant's female reproductive part, called the pistil. Water lilies, for example, which are closed in the morning and evening, open during midday when the weather is warmer and insects are more active. It is usually at this time that the insects pick up the pollen and transfer it from the stamen to the pistil, allowing pollination to occur.

Some wetland plants send out rhizomes, which are stems that spread out under water or soil and form new plants. Reed mace and common reeds are examples.

Common green plants Common wetland green plants include mangrove trees; insectivorous plants; and reeds, rushes, and sedges.

MANGROVE TREES Trees that live in wetlands must tolerate having their roots wet for a long period of time. One of the best examples of this is the mangrove tree, which grows in tropical and subtropical saltwater swamps. Mangroves have done so well at adapting to wetland conditions that there are more than 34,000,000 acres (14,000,000 hectares) of them in the world. One of the largest mangrove forests is the Sundarban Forest in Bangladesh. Mangroves can also be found in southern Florida and other tropical areas.

Mangroves have special membranes that stop or reduce the entry of salt into their systems. Because they stand in waterlogged soil, mangroves get their oxygen from pores in their roots, the largest portions of which are above ground, and from small gaps in their bark.

Mangrove flowers are pollinated by the wind, and their seedlings begin to grow while they are still attached to the parent tree. Seedlings germinate (grow) within the fruit that contains them. The roots of the maturing seedling break through the outer wall of the fruit, appearing as many tendrils. When the seedling reaches a mature enough stage, it is released from the parent plant, floats on the water, and then settles in mud, where it establishes roots and begins to grow.

INSECTIVOROUS PLANTS Two common peatland plants eat insects. The sundew eats ants by catching them in a sticky liquid on its leaves. It then releases chemicals that break down the ant's body so the plant can absorb the nutrients. The pitcher plant gets its name from the shape of its leaf, which looks like a pitcher. If an insect crawls down the leaf it cannot crawl back up because the leaves are too slippery. It is then digested by chemicals in the plant.

Mangrove trees are common wetland plants. (Reproduced by permission of J L M Visuals.)

REEDS, RUSHES, AND SEDGES A reed is a type of grass with tall, feathery flowers; round, hollow stems; and long, flat, narrow straplike leaves. Although the rush is grasslike, it is not a true grass. It has round, solid stems

and narrow, rigid leaves. Sedges are also grasslike but are not true grasses. They have solid, triangular stems.

Reed mace and common reeds are two plants that play a key role in wetland succession. They reproduce widely through their rhizomes. As the plants spread out, they push aside other plants and quickly choke up a pond, slowing down the water's movement and trapping dead plant matter. As this matter accumulates, the pond fills up and becomes a marsh.

ENDANGERED SPECIES

Twenty-six percent of the endangered plants in North America are dependent upon wetlands for their survival. Environmentalists predict that about 400 American plants will be extinct in the near future. At the same time, endangered plants will number almost 700. (When no living members survive, a species is extinct. A species that is threatened with the loss of many members and may one day become extinct is endangered.)

Changes in the habitat, such as succession, pollution, and new weather patterns, endanger wetland plants. They are also endangered by people who collect them. Rare species of orchids are an example, as is the Venus' flytrap which has become a popular houseplant and has been overharvested. Peat moss is also overharvested because it is popular as a fuel, as a packaging material, and in gardens.

An insectivorous pitcher plant commonly found in peatlands. (Reproduced by permission of Field Mark Publications. Photograph by Robert J. Huffman.)

ANIMAL LIFE

Wetlands have been called "biological supermarkets." Besides animals that live there permanently, many nonwetland animals, such as opossums, raccoons, and skunks, visit for food and water. Wetlands also provide shelter for mammals such as minks, moose, and muskrats. Wetland conditions make it necessary for the animals that live there permanently to adapt in special ways.

MICROORGANISMS

Microorganisms cannot be seen by the human eye. Those found in wetlands include protozoa and bacteria. Protozoa are single-celled animals that can use the Sun's energy to make their own food. Other protozoa are parasites that live on aquatic plants, on damp ground, or inside animals or plants. Bacteria are also single celled, but they cannot make their own food and must obtain it from the environment.

Some protozoa infect mosquitoes with malaria, a disease characterized by chills, fever, and sweating. The mosquitoes then spread the disease to humans. Malaria is found in temperate, subtropical, and tropical regions. Although its spread has been greatly reduced due to the use of pesticides to kill mosquitoes, it is still a problem in parts of Africa and in southeastern Asia.

INVERTEBRATES

Invertebrates are animals without a backbone. They range from simple sponges to complex animals such as insects, snails, and crabs.

More than twenty species of insects have adapted to life in the wetland. Some spend their entire lives in the water. Others live in the water while young but leave it when they become adults. Still other insects have gills, just like fish, which enable them to obtain oxygen from the water; and some breathe air. For example, the diving beetle traps air in the hairs on its body or under its wings, which helps it to float. Water spiders create bubbles of air under the water in which they feed and lay eggs. The water spider even hibernates underwater. When the weather gets cold, it constructs an air bubble in deep water and remains there until spring.

Some invertebrates, such as crabs, are able to survive in wetlands by creating watertight holes in the mud or sand and hide there during high tide. Other species, like the African mangrove snail, feed in the mud and sand when the tides are low. When the tides come in the snail climbs the mangrove trees to keep from drowning.

Food Insects may feed underwater as well as on the surface. For instance the milkweed beetle feeds on insects that can be found on wetland plants, while the diving beetle adds tadpoles and small fish to its diet. The dragonfly larva eats tiny floating organisms and water fleas. The larva of a certain species of caddisfly weaves a silk net under the water in which to catch floating algae.

Some snails are plant feeders, and others eat the eggs and larvae of other invertebrates, or even decaying matter. Crabs are often omnivorous, meaning they eat both plants and animals.

Reproduction Most invertebrates have a four-part life cycle that increases their ability to survive in wetlands. The first stage is the egg. The second

ATTACK OF THE KILLER PLANT

Insectivorous (insect-eating) plants, such as the sundew, the pitcher plant, and the Venus' flytrap, need nitrogen to survive. Because these plants live in bog areas where the soil is acidic, nitrogen is not readily available. In order to obtain the proper nutrition, they trap insects, dissolve them, and absorb their nutrients, which include the essential nitrogen.

The Venus' flytrap, for example, has toothed jaws at the end of its leaves. These jaws spring shut when an insect looking for nectar touches the sensitive hairs found inside. The plant then eats the insect by dissolving it with digestive juices and absorbing it through its leaves. After the insect has been digested, the plant opens its jaws again and waits for the next victim.

Venus' flytraps are found in the marshy coasts of North and South Carolina. However, because they grow only under certain conditions and because they are popular houseplants, they have been overharvested and are endangered.

stage is the larva, which may actually be divided into several steps where the insect increases in size, between which there is a shedding of the outer casing or skin. The third stage is the pupal stage, during which the insect lives in yet another protective casing. Finally, the adult emerges. Some wetland dragonflies lay their eggs in the tissues of submerged plants. The insects remain in the water as larvae. As they mature, they move to dry land.

Common invertebrates Perhaps the most well-known and unpopular wetland invertebrate is the mosquito. Mosquitoes breed in shallow wetland waters. In their larval form, they have tubes in their abdomens that stick out of the water, allowing them to breathe. Even though the larvae are often eaten by frogs, fish, and aquatic insects, many mosquitoes survive. Those that do face being eaten by bats, sprayed with pesticides, and swatted by humans. In spite of these hazardous conditions, mosquitoes in large numbers are responsible for diseases in both humans and animals.

AMPHIBIANS

More than 190 species of amphibians can be found in wetlands. Amphibians are vertebrates, which means they have a backbone. There are two kinds of amphibians, those with tails, like salamanders and newts, and those without tails, like toads and frogs.

Amphibians live at least part of their lives in water and are found in primarily freshwater environments. Most are found in warm, moist regions and in a few temperate zones. Because amphibians breathe through their skin, they must usually be close to water so they can stay moist. Only moist skin can absorb oxygen. If they are dry for too long they will die. Their lower layer of skin, called the dermis, also helps them to stay moist by producing mucous, a thick fluid that moistens and protects the body.

Amphibians are cold-blooded animals, which means their body temperatures are about the same temperature as their environment. They need warmth and energy from the Sun in order to be active. As temperatures grow cooler, they slow down and seek shelter so they can be comfortable. In cold or temperate regions, some amphibians hibernate (become inactive) during the winter, hiding in mud or trees. When the weather gets too hot, they go through another period of inactivity called estivation.

THE MARSH THAT WOULDN'T GIVE UP

For more than 2,000 years, popes, emperors, and government officials tried to drain the Pontine Marshes, where malaria-carrying mosquitoes like to live. Located in south central Italy, the marshes are bordered by the Tyrrhenian Sea, the Lepini Mountains, and the Alban Hills.

All attempts to completely drain the marshes failed. The area survived and about 300 square miles (777 square kilometers) of the original marsh is now a national park with native vegetation intact. With the help of water from the marsh, the area is now one of the most agriculturally productive in Italy. Crops grown in the area include sugar beets, fruits, grains, and vegetables. Livestock is also raised here.

Food In their larval form, amphibians are usually herbivorous (plant eating). Adult amphibians are usually carnivorous (meat eating), feeding on insects, slugs, and worms. Salamanders that live in the water suck their prey into their mouths. Those that live on land have long, sticky tongues that capture food. One of the favorite foods of the frog is the mosquito.

Reproduction Most frogs, newts, and toads lay lots of tiny eggs. Some are held together in a jelly-like substance. As the female lays her eggs in the water, the male releases sperm, which is carried to the eggs by the water. The female newt, for example, lays from 200 to 400 eggs on the submerged leaves of aquatic plants to await fertilization from the male's sperm.

Common amphibians The most common wetland amphibians are salamanders, frogs, and toads. Frogs and toads can be found all over the world, at all altitudes, and in both fresh and salt water.

FROGS The frog spends half its life in the water and half out of it. Frogs can lay up to as many as 3,000 eggs, which float beneath the water's surface. Frog eggs hatch into tadpoles (larvae), which swim and breathe through gills, like those on a fish. The larvae feed on small plants and animals in the water. As they mature, they develop legs and lungs, which they will need on land.

Adult frogs feed mainly on insects, especially mosquitoes, but bullfrogs have been known to eat birds and snakes. In the food web, they in turn become meals for herons, raccoons, and other wetland animals.

Frogs are considered "bio-indicators." This means that when many frogs are sick their entire environment may be in trouble. If they disappear from a habitat or if they do not sing, the area may be polluted and its resources may be dwindling. Frogs are totally absent from some wetlands in the United States, indicating their environment was no longer a healthy place for them to live.

REPTILES

Reptiles are also cold-blooded vertebrates that depend on the environment for warmth.

THE BUG THAT WALKS ON WATER

The marsh treader can be found walking very slowly across the surface of marsh water. It has a long, skinny body, about .3 inch (8 millimeters) in length. Its legs are as fine as threads, and its body is covered with velvety hairs. As it moves across the water, it searches for food. Mosquito larvae are its favorite dish.

THE WEB OF LIFE: MORPHOGENIC FIELDS

Rupert Sheldrake, a British biochemist, has a new idea about evolution (how the characteristics of a species change over time in response to the environment). Sheldrake believes that all living things are influenced by these invisible "morphogenic fields." The form and behavior of other organisms of the same species that lived in the past create these invisible fields.

For example, the monarch butterfly travels more than 2,000 miles (3,218 kilometers) from its summer home in the United States to its winter home in Mexico. While they are in their southern homes, the butterflies breed and then die. However, the next generation of butterflies knows how to migrate north again in the spring, although no living member of the past generation is there to show them the way. They seem to have mysteriously acquired the knowledge.

Thousands of species of reptiles live in the temperate and tropical wetlands of the world. They include snakes, lizards, turtles, and crocodiles. Reptiles are more active when the weather and water temperature become warmer. Unlike amphibians, though, reptiles have skin that are waterproof, and therefore they do not dry out. This allows them more freedom to move away from the wet areas.

Because they are so sensitive to their environment, reptiles often go through a period of hibernation in cold weather. Turtles bury themselves in the wetland mud. They barely breathe, and their energy comes from stored body fat. When the weather warms in spring, they come out of hibernation and become active. At the other extreme, when the weather becomes very hot and dry, some reptiles go through estivation, another inactive period.

Food Some wetland reptiles, such as the dice snake, are carnivores, eating frogs, small fish, and crayfish. The pond turtle, which inhabits tropical swamps, is an omnivore. When it is young it feeds on insects, crustaceans, mollusks, and tadpoles. As an adult, it eats primarily wetland plants.

> ### THE SALAMANDER THAT NEVER GROWS UP
>
> The axolotl, a type of salamander found in freshwater wetlands in Mexico and the western United States, spends its entire life as a tadpole. Although it can reach a size of 10 inches (25 centimeters) and is able to reproduce, it never matures, so it never leaves the water. Scientists have determined that the diet of these salamanders is lacking the chemical iodine, which is in short supply in many parts of the world. In fact, this shortage is the reason why iodized table salt is used in most homes. Iodine has been added as an essential nutrient.

Reproduction The eggs of lizards, alligators, and turtles are either hard or rubbery and do not dry out easily. Most are buried in the warm ground, which helps them hatch. In a few species of lizards and snakes the young develop inside the female's body until birth.

The dice snake lays it eggs on land, close to the water. Crocodiles and alligators keep eggs in warm nests, which can be simple holes in the ground or constructions above the ground made from leaves and branches.

Common reptiles Two well-known and dangerous wetland reptiles are the crocodile and the cottonmouth snake.

CROCODILE Crocodiles are found in warmer parts of North and South America, Africa, Australia, and Southeast Asia. They have inhabited the Earth for more than 170,000,000 years. Perhaps their long survival has been due in part to the fact that they will eat almost anything or anyone. Crocodiles are responsible for killing at least 10 humans every day in Africa, and a large crocodile will attack even a 4,000-pound (1,816-kilogram) adult rhinoceros if given the chance. A crocodile's teeth are rounded and made for holding, not cutting, and it first tries to swallow its prey whole. If the prey is too large, the crocodile pulls it underwater and stashes it where it can decay. When the meat is soft, the crocodile rips off a piece at a time.

The largest crocodile officially recorded was almost 20 feet (6 meters) long, more than 4 feet (1.2 meters) tall when upright, and probably weighed well over a ton (.9 metric ton). A crocodile continues to grow throughout its life, and a very large one may be as much as 200 years old. Their speed out of the water over short distances may be more than 35 miles (56 kilometers) per hour. By contrast, a very fast human may run 25 miles (40 kilometers) per hour over the same distance.

Frog embryos complete their metamorphoses inside of eggs. (Reproduced by permission of Corbis. Photograph by Michael and Patricia Fogden.)

COTTONMOUTH SNAKE The cottonmouth snake inhabits marshes and swamps in the south and southeastern United States. A thick-bodied snake, it spends most of its life in or near water. It is active at night, when it preys on amphibians, fish, snakes, and birds. Its venom is extremely poisonous. Because the venom has the ability to coagulate (solidify or partially solidify) blood, it is extracted and used medically in treating bleeding disorders.

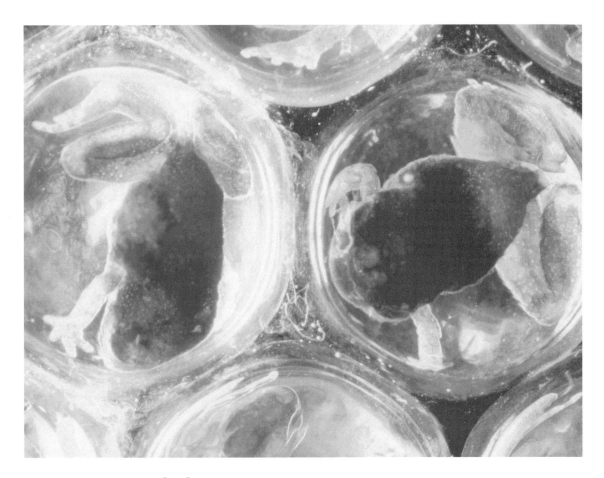

FISH

Like amphibians and reptiles, fish are cold-blooded vertebrates. They use fins for swimming and gills for breathing. Two-thirds of all fish used for human food depend on wetlands during some part of their lives. Most commercial game fish breed and raise their young in marshes and estuaries. When the wetlands are not healthy, the fish die, and the commercial and recreational fishing industries suffer.

Food Some fish eat plants while others depend upon insects, worms, crustaceans, or smaller fish. Some fish may simply grab an insect from the surface of the water or use a more elaborate scheme. The archerfish, for example, has grooves in the top of its mouth that lets it spit water at its prey. The force of the spit knocks the insect off its perch and into the water, where it is quickly eaten. Archerfish are found in the swamps of southeast Asia.

Reproduction Fish that breed in wetlands include the flounder, sea trout, striped bass, and carp. Some European carp live in rivers and move to the floodplains during spawning (breeding) season. Others live in the ocean and spawn in mudflats, marshes, or mangrove swamps.

Common fish Fish that commonly rely on coastal wetlands are striped bass, sea trout, African lungfish, mudskippers, and flounder. Brackish marshes, which contain both salt and fresh water, support sole, sardines, and common mackerel.

AFRICAN LUNGFISH The African lungfish, which lives in pools surrounded by swampland, has adapted to the wet/dry cycle of wetlands. When there is plenty of water, the fish breathes through its gills. In the dry season, when the water disappears, the fish burrows into the mud and seals itself into a moist cocoon, breathing through special pouches on its underside. It can remain inactive in its cocoon for months or even years.

MUDSKIPPER Also well adapted to life in the wetland is the mudskipper, found in coastal areas of Bangladesh. This fish usually lives out of the water on exposed mud at low tides. Mudskippers breathe air through membranes at the back of their throats. Enlarged gill chambers hold a lot of water, which helps them remain on land for long periods. They also keep water in their mouths, which they swish over their eyes and skin to help them stay wet. They feed on crabs that are easily caught during low tide.

> ## THE WEB OF LIFE: KEYSTONE SPECIES
>
> Species interact with one another in many ways. Those that play an important role in the survival of others are called "keystone species." In the wetlands, one keystone species is the alligator, which lives in southern marshes such as the Florida Everglades.
>
> Wetlands often dry up if rainfall is low. The alligators then dig holes with their powerful tails and bodies. Water collects in these holes and helps the alligator, as well as birds, fish, frogs, turtles, and other marsh animals, to survive during dry periods. However, it is often risky for other animals to drink at an alligator's hole—for obvious reasons.

BIRDS

Many different species of birds live in or near wetlands. These include many varieties of wading birds, waterfowl, shore birds, and perching birds. One of the special adaptations of birds to wetlands is their bill. Some bills, like those of the egret, are shaped like daggers for stabbing prey such as frogs and fish. Other bills, like those of the spoonbill, are designed to root through mud in search of food.

Wetlands provide a variety of hiding places in which birds can build nests and protect their young. Dense underbrush or hollows in the ground are good hiding places, especially when nests are made from reeds, flowers, and grasses, which help them blend. Wood ducks build their nests in hollow cavities in the trees that line the wetland shores.

Food Plants and small animals in wetlands provide a ready food source for birds. Some birds feed on vegetation, while others are predatory. The pike and marsh harrier, for example, feed on mussels, small fish, or insects and their larvae. Herons and egrets feed on fish, frogs, and snakes.

Reproduction All birds reproduce by laying eggs. Male birds are brightly colored and sing, both of which attract the attention of females. After mating, female birds lay their eggs in nests made out of many different materials. These nests may be found in a variety of places throughout the wetland area. Different species of birds lay varying numbers of eggs. The tufted duck, for example, lays from six to fourteen eggs, which hatch in a little less than a month. The baby ducks begin to swim within days.

Common birds Birds found in wetland environments can be grouped as wading birds, shorebirds, waterfowl, and perching birds.

WADING BIRDS Wading birds, such as herons and egrets, have long legs for wading through the shallow water. They have wide feet, long necks, and long bills that are used for nabbing fish, snakes, and other food. Herons and egrets are the most common in freshwater marshes of North America. The great blue heron stands 4 feet (1.2 meters) tall. This is the tallest recorded wading bird in North America.

SHOREBIRDS Shorebirds feed or nest along the banks of wetlands and prefer shallow water. Their feet are adapted for moving in water, and some have long, widely spread toes to prevent them from sinking in the mud.

THE JERSEY DEVIL

The Jersey Devil is a mythical creature of the New Jersey Pine Barrens that has haunted New Jersey residents for almost 400 years. It has seemed so real to some people that it has caused entire towns to live in fear, and factories and schools to close. Some say the Jersey Devil is just a legend, while others say there is evidence to support the existence of an animal or supernatural being.

According to folklore, one person who saw the Jersey Devil and lived to tell about it said it was more than 3 feet tall, with a head like a collie dog and a face like a horse. It had a long neck, wings about 2 feet long, back legs like those of a crane, and the hooves of a horse. "My wife and I were scared, I tell you," the man said, "but I managed to open the window and say, 'Shoo,' and it turned around, barked at me, and flew away."

The Jersey Devil became the state's "official demon" in the 1930s. Some believe that the Jersey Devil is really a sand hill crane, but the mystery lives on.

The bills of shorebirds are adapted to help them find food. The ruddy turnstone, for example, has a short, flattened, upturned bill, which helps it sift through mud or overturn pebbles and shells. A favorite food of the oystercatcher is the mussel. Each young oystercatcher learns from its parents a technique for opening the mussel shells to get at the meat. Some birds hammer a hole in the shell with their bills, and others prop the shells at an angle so they can pry them open. It may take the birds several years of practice to get the technique just right.

The more than 200 species of shorebirds inhabiting wetlands include sandpipers, which are found in marshes, wet woodlands, and on inland ponds, lakes, and rivers.

WATERFOWL Waterfowl are birds that spend most of the time on water, especially swimming birds such as ducks, geese, and swans. Their legs are closer to the rear of their bodies than those of most birds, which is good for swimming, but awkward for walking. Their bills are designed for grabbing wetland vegetation, such as sedges and grasses, on which they feed.

A great blue heron stands with a fresh fish in its bill for its next meal. (Reproduced by permission of Corbis. Photograph by Kennan Ward.)

PERCHING BIRDS Perching birds can also be found living along wetland areas where food and shelter are readily available. They are land birds, with feet designed for perching. Their feet usually have three long toes in the front and one in the back. The barred owl can be commonly found in swamps, while red-winged blackbirds live in cattail marshes, and bogs are home to golden plovers, skylarks, and meadow pipits.

THE DUCK THAT SAVED WETLANDS

The Federal Duck Stamp Program began in 1934 when the U.S. Congress passed the Migratory Bird Hunting Stamp Act. The act required waterfowl hunters to purchase and carry Federal Duck Stamps, which are attached to a hunting license or other document. The money was used to buy or lease waterfowl habitats from their owners and designate them as restricted areas, helping to save millions of acres of wetland.

Over the years, efforts have also included saving wetland species that were declining in numbers, such as wood ducks, canvasback ducks, and pintail ducks. The program also helps endangered species that rely on wetlands for food and shelter.

Today, anyone can own Federal Duck Stamps. State, international, and junior stamps are also available. Anyone having a duck stamp gets free admission to all National Wildlife Refuges where entrance fees are charged.

To learn more about the duck stamps and to see pictures of them, visit http://www.nationalwildlife.com

or contact:

The Environmental Protection Agency Office of Wetlands, Oceans, and Watersheds
Wetlands Division (4502F)
401 M Street SW
Washington, DC 20460

MAMMALS

Mammals are warm-blooded vertebrates that are covered with at least some hair and that bear live young. Aquatic mammals, such as muskrats, have waterproof fur that helps them blend into their surroundings and webbed toes for better swimming. Some of these mammals live permanently in wetlands and others, such as raccoons, visit for food, water, and shelter during some part of their lives.

Food Some aquatic mammals, like the otter, are carnivores, eating rabbits, birds, and fish. Muskrats, however, are omnivores, eating both animals, such as mussels, and plants, such as cattails. Beavers are herbivorous and eat trees, weeds, and other plants.

Reproduction Mammals give birth to live young that have developed inside the female's body. Some mammals are helpless at birth while others are able to walk and even run immediately. Some are born with fur and with their eyes and ears open. Others, like the muskrat, are born hairless and blind. After about three weeks, however, young muskrats are able to see and swim.

Common mammals The American beaver is well-adapted to the wetland environment. It has webbed feet for powerful swimming and warm, waterproof fur. Other mammals that can be found in wetlands include the mink, the water shrew, and the Australian platypus. The sitatunga, a type of antelope, lives near swamps in central and east Africa. It feeds on emergent wetland plants.

RED DEER Although peatlands generally do not support many species of animals, the red deer at 4 feet (1.2 meters) tall—the largest animal found in Ireland—lives in the bogs. It can be seen rolling in the peat in order to get rid of parasites and insects.

MUSKRAT The muskrat, which resembles a beaver, is a heavy-bodied rodent about 12 inches (30 centimeters) long, not including the long tail. It is a native of North America and common to marshes all over the country. Its hind feet are webbed for swimming and it can often be seen floating on the water's surface. Marsh plants such as sedges, reeds, and the roots of water plants provide most of its diet.

Muskrats build dome-shaped houses in water. They pile up mud, cattails, and other plants until the mound rises above the water's surface. Tunnels lead into the mound in which one or more rooms are hollowed out above the water's surface.

Muskrats have been hunted for their fur, which is brown and consists of soft underfur and a dense coat. They are also sold as food, often labeled as "marsh rabbit."

> ### THE RABBIT THAT SWIMS LIKE A DUCK
>
> The North American swamp rabbit has broad, flat feet so that it can move easily over wet, marshy soil. It can also swim, and often escapes danger by staying submerged with only the tip of its nose exposed to the air in order to breathe.

ENDANGERED SPECIES

Over half of the endangered or threatened fish and other wildlife in the United States—about 240 species altogether—rely partly on wetlands for food, water, shelter, or a place to reproduce. Over one-third of these live only in wetlands.

In North America, endangered species include the whooping crane and the manatee. The whooping crane lives in coastal swamps and feeds on roots and small reptiles. It has been threatened by hunting, pollution, and dredging (dragging a net along the bottom of a body of water to gather shellfish or plant specimens).

The manatee is a slow-moving, seal-like animal. Manatees live in shallow coastal wetlands in the Caribbean, the Amazon, and Africa. They have become endangered because of overhunting, being caught and strangled in fishing nets, or being killed by boat propellers. Pollution has also affected their habitats and food sources.

In Ireland, the Greenland white-fronted goose, which relies on bogs for feeding and breeding, is endangered because many bogs have been lost.

HUMAN LIFE

After 13000 B.C., wetlands played an important role in early civilizations. For prehistoric communities throughout the world they provided food, water, and materials for clothing, shelter, and tools. Wetlands are able to reveal many of these ancient civilizations' secrets to scientists. Bodies of animals and humans, and artifacts (objects made by humans, including tools, weapons, jars, and clothing) have been well preserved in some wetlands and give information about how ancient people lived. Evidence of these ancient communities has been found in bogs. Conditions in bogs are excellent for preserving artifacts and bodies.

IMPACT OF WETLANDS ON HUMAN LIFE

Wetlands have had an affect on the supply of food and water, on shelter, and on other resources.

Food and water Wetlands are a source of water for drinking and crops. Areas near wetlands also continue to provide food in a variety of ways for tribal cultures in Asia and Africa and for urban peoples elsewhere. Since wetlands are a habitat for much wildlife, hunting and fishing are common there.

Cattle can be raised near wetlands because water and grazing land are available. Crops such as sorghum (a cereal grain native to Africa and Asia) are commonly grown in wetland areas. Grain sorghums, such as milo, kafir, and durra, have adapted to the extremes of the wet/dry cycle of the wetland and are among the most drought-tolerant grains. Millions of people in China,

---

THE BOG PEOPLE

The organic materials in peat bogs absorb and hold large amounts of water like a sponge. Bog soils are extremely acidic and have little oxygen. Because of these factors, decomposition takes place very slowly. Entire trees, animals, and even human bodies have been preserved for centuries.

The remains of more than 2,000 humans have been found in bogs, primarily in northwestern Europe. The bodies are in various stages of preservation, from skeletons to those with flesh intact. Sometimes only body parts such as heads or limbs are found. Often the skin has darkened and the hair has turned red from peat acids. The bodies found in Europe range in age from about 1,500 to 3,000 years old.

Many of these people appear to have died violent deaths. Perhaps they were killed as a punishment for a crime or were the victims of human sacrifice. Their deaths may have taken place at the bog because it was so isolated.

At a bog in Windover, Florida, 160 bodies have been found that are about 7,000 years old. Buried with them are food, tools, clothes, and weapons. Many of these items were made from wetland plants and from the bones of wetland animals.

For further information, visit the official on-line publication of the Archeology Institute of America:
http://www.archaeology.org/online/features/bog/ or

India, and Africa rely on sorghum as a food staple. In the United States, sorghum is used mostly as livestock feed.

Besides sorghum, American farmers use wetland plants, such as marsh grass, reeds, and sedges, for feeding livestock. Wetlands are also an excellent place to grow rice, which is a major food source for much of the world's population.

Shelter Materials for dwellings are also available from wetland sources. Roofs made from reeds are used on huts in Egypt and in stilt houses in Indonesia to keep the occupants cool and dry. A tightly woven reed roof can last for 40 years. House frames are constructed from the timber of mangrove, palm, and papyrus trees. Wetland sediments, such as clay and mud, may be used to produce bricks that are used for walls.

Other resources Wetlands provide other resources that humans rely on besides building materials. For example, dried peat is used in homes for heat and to fuel electric generators in countries, such as Ireland, where coal is scarce. Peat is a source of protein for livestock feed, and chemically processed peat is used as a base for polishes and waxes. In horticulture, peat moss is used as groundcover, a soil conditioner, and a growing medium. Peat is also found in medicinal baths and cosmetics.

> ### SAVING THE LANDSCAPE WITH CREATIVE ARCHITECTURE
> Seattle architect James Cutler believes that the environment should be preserved when designing and building houses. He goes to any extreme to save the landscape, including building houses on stilts and placing sidewalks over the tops of forests. When Cutler designed a home for Microsoft chief Bill Gates, Cutler reestablished wetlands on the property.

Other wetland soils are sources of gravel and phosphate. Phosphate is used as a raw material for making fertilizers, chemicals, and other commercial products.

About 25 percent of medicinal drugs come from plants, some of which grow in wetlands. The Madagascar periwinkle, for example, provides vineristine, an agent that has helped reduce the death rate of childhood leukemia.

IMPACT OF HUMANS ON WETLANDS

The important role of the wetland for plant and animal communities and for the environment in general has not always been understood. For more than 2,000 years, people in different countries have been draining wetlands to get rid of mosquitoes and disease, to increase land for farming, and to make room for development. By 1990, more than half of the wetlands in the United States were destroyed, and more than 90 percent of bogs in the United Kingdom are gone.

As the value of wetlands becomes more recognized, government agencies, such as the U.S. Department of the Interior and U.S. Fish and Wildlife Service, as well as environmental groups, are working to preserve existing wetlands and to create new ones.

The Convention on Internationally Important Wetlands was signed by representatives of many nations in Ramsar, Iran, in 1971. Commonly known as the Ramsar Convention, it is an intergovernmental treaty that provides the framework for national action and international cooperation for the conservation and wise use of wetlands and their resources.

Use of plants and animals When animals and plants are overharvested they become endangered. Overharvesting means they are used up and destroyed at a faster rate than they can reproduce. When this happens, wetland, timber, fuel, medicines, and sources of food for humans and animals are all lost.

Use of natural resources The two primary resources found in wetlands are water and peat. Wetlands often act as part of the groundwater system. When they are lost, the supply of drinking water may be affected.

> ### ADOPT A WETLAND
> Both the federal government and individual states are working to protect wetlands. Some states sponsor "Adopt a Wetland" programs, in which groups of people agree to help support their local wetlands in a variety of ways, such as picking up litter. For information about adopting a wetland, contact the Environmental Protection Agency, Public Information Center, (202) 260-7756 or (202) 260-2080.

It takes twenty years for one inch of peat to form, and its overuse has caused significant peatland losses. Western European peat mining companies are rapidly using up local areas of peat and expanding their mining operations into eastern European countries. All natural peatlands or bogs in the Netherlands and Poland have been destroyed. Switzerland and Germany have only 1,250 acres (500 hectares) each remaining.

Quality of the environment Wetlands are endangered by industrial and municipal (city) contaminants, by the accumulation of toxic chemicals, and from acid rain.

Acid rain is a type of air pollution especially dangerous to wetlands. It forms when industrial pollutants such as sulfur or nitrogen combine with moisture in the atmosphere and form sulfuric or nitric acids. These acids can be carried long distances by the wind before they fall either as dry deposits or in the form of rain or snow. Acid rain can significantly damage both plant and animal life. It is especially devastating to wetland amphibians such as salamanders, because it prevents their eggs from maturing.

Mining for minerals near a wetland can have a negative impact on the biome. Mercury, a poisonous liquid metal used in gold mining operations, often contaminates wetlands close to the mines. Mining operations also require a lot of water, and nearby wetlands may be drained.

The world's climate may be growing warmer because of human activity. If polar ice caps melt, there will be a rise in sea level. As this happens, more salt water will flood into coastal wetlands and increase the salinity not only of the wetlands but of rivers, bays, and water supplies beneath the ground. As these habitats change, animal and plant life would be affected, and some wetlands may be destroyed.

Artificial wetlands Artificial wetlands are those created by humans. In 1989, U.S. president George Bush, asked that the United States work toward the goal of no net loss of wetlands. This means that if a natural wetland is destroyed by development, an artificial one must be built to replace it. However, creating a wetland is very difficult. All the right conditions need to be met, including a water source and soil that contains a lot of water but not much oxygen.

An artificial wetland that works quite well and is extremely valuable worldwide is a rice paddy. A rice paddy is simply a field that is flooded for the purpose of growing rice, a food staple for about 3,000,000,000 people—more than one half of the world's population. Some are flooded naturally, by monsoon (tropical) rains or overflowing rivers. Others are flooded by irrigation (watering). Mud dams and waterwheels are built to bring in and hold the water level at approximately 4 to 6 inches (10 to 15 centimeters) while the rice grows. Ninety percent of rice paddies are in Asia, especially China and India. However, some rice is also grown in Europe and the United States.

Artificial wetlands can also be found on the Arabian Peninsula, where they are used for water storage and sewage treatment. Other human-made wetlands have been created as winter homes for migrating waterfowl.

NATIVE PEOPLES

The Marsh Arabs of Iraq and the Nilotic peoples of Africa live in or near wetlands that provide almost all of the resources they need.

Marsh Arabs of Iraq The Mesopotamian wetland, one of the largest in the world, lies between the Tigris and Euphrates Rivers in southern Iraq. The Marsh Arabs, or Ma'dan, have lived there for more than 6,000 years. The water there is clean, calm, and fairly shallow—about 8 feet (2.4 meters) deep.

MARSHES CLEAN UP MESSES

A lot of contaminants seep into the environment every day from oil spills, sewage treatment plants, industrial waste, and old mines. In some areas, phytoremediation (fi-toh-rih-mee-dee-AY-shun) is being used to clean up these messes. (Phyto is a Greek word for plant and remediation means to make things right.) In phytoremediation, plants are used to absorb pollutants.

One type of phytoremediation involves creating an artificial marsh containing cattails and water lilies. These wetland plants are able to absorb some water pollutants. Poplar trees can clean up water polluted with oil, and sunflowers are used to help clean up radioactive materials in the soil. As a matter of fact, sunflowers are being used near the Chernobyl nuclear power station in Pryp'yat near Kiev in the Ukraine. The power station exploded in 1986, releasing radioactive waste into the environment.

HOUSES WITHOUT WALLS

The Seminoles, a Native American people who once lived in the Florida Everglades, built homes with no walls. Called a *chikee,* the home was built on a platform and had a log framework and a thatched reed roof. The absence of walls provided lots of ventilation in the warm climate.

The Marsh Arabs' houses are made of reeds and built on islands that they also make. To construct the islands, they create a fence from reeds and partially submerge it. Then they fill the area inside the fence with cut rushes, add layers of mud, and stamp it all down. When the pile reaches the water's surface, they fold the top of the fence onto the pile and add more reeds to finish it. The completed island is big enough not only for the family to build a house and live on, but for their cattle as well.

Because their home is isolated from the outside world, the Marsh Arabs' way of life has remained the same for hundreds of years. Daily life consists of fishing, buffalo herding, and growing rice. Reeds are gathered every day to feed the buffalo. Transportation from house to house or to other villages is by means of small canoes called *mashhufs*.

The survival of the Marsh Arabs is threatened, however, by irrigation practices, which have drawn water from their wetlands. Some of the marshes have already been drained for agricultural use and oil exploration.

Nilotic peoples The Nilotic peoples live near the Nile River or in the Nile Valley in Africa. Two Nilotic tribes, the Dinka and the Nuer, live in southern Sudan. This area is a rich floodplain, and during the wet season, from July through October, the people live on high ground in permanent villages. From December through April, when the floodwater recedes, they move to the floodplains. There, the Dinka build temporary villages right along the banks of the Nile River. Their homes, made from reeds and grasses, are circular at the base and come to a point at the top, much like a teepee. Their neighboring tribe, the Nuer, lives in a similar fashion but in the marsh, savanna (grassland), and swamp areas.

Wetland grasses and plants grow in the floodplain after the waters recede. The grasses provide food for cattle and also attract wildlife, which is often hunted for food. Nilotic tribes fish, hunt, and grow some grains. The Dinka use their cattle for meat and milk, but the Nuer will eat cattle only during religious ceremonies. For them, cattle represent wealth and are used for bridal dowries (gifts) and to sell. Animal hides are made into clothing and bedding, and dried cattle dung (waste) is used for fuel.

A Marsh Arab father and daughter in traditional dress. These people rely on wetlands to provide almost all of the resources they need to live. (Reproduced by permission of Corbis. Photograph by Nik Wheeler.)

THE FOOD WEB

The transfer of energy from organism to organism forms a series called a food chain. All the possible feeding relationships that exist in a biome make up its food web. In the wetland, as elsewhere, the food web consists of producers, consumers, and decomposers. These three types of organisms transfer energy within the wetland environment.

At the bottom of the food chain are the producers, which may be single-celled organisms such as bacteria and protozoa. The shallow wetland water lets in a lot of sunlight, which helps these organisms grow.

These primary producers are eaten by primary consumers such as the larval forms of frogs and toads, and larger animals such as shrimp and snails. Other primary consumers, such as small aquatic insects, shellfish, and small fish, feed on plant materials.

Primary consumers, in turn, are food for predators, such as larger fish, reptiles, amphibians, birds, and mammals, which are called secondary consumers. Tertiary consumers are the predators, such as owls, coyotes, and humans, that prey on both primary and secondary consumers.

A Marsh Arab village. The houses are made of reeds and are built on small islands. (Reproduced by permission of Corbis. Photograph by Nik Wheeler.)

The decomposers, which feed on dead organic matter, include fungi, bacteria, and crabs.

SPOTLIGHT ON WETLANDS

THE FLORIDA EVERGLADES

The Florida Everglades, lying within the tropical zone near the equator, is one of the largest freshwater marshes in the world, stretching from Lake Okeechobee 100 miles (160 kilometers) south to Florida Bay. Within its boundaries are also freshwater swamps and coastal (saltwater) mangrove swamps. In the northern portion of the Everglades is Big Cypress Swamp. This swamp covers 1,500 square miles (4,000 square kilometers) and gets its name from the many tall bald cypress trees that live there.

The Everglades were formed during the last Ice Age, which ended about 10,000 years ago. During this period, glaciers melted and sea level was raised, which flooded the area and turned it into a wetland. Under the water and soil is porous limestone rock formed during the last glacial period. This type of rock contains shells and skeletons of animals deposited in the sea. Hummocks (rounded hills or ridges) have formed in some areas, and there hardwood trees grow. Otherwise, the terrain is flat.

The subtropical climate of the Everglades is mild, with temperatures ranging from 73° to 95°F (25° to 35°C) in the summer. Winters are mild, with an average high of 77°F (25°C) and a low of 53°F (12°C). The rains occur from June to October, with an annual accumulation of about 55 inches (1,400 millimeters). The region is often hit by severe tropical storms.

The grassy waters of the Everglades are actually very shallow, ranging from about 6 inches (15 centimeters) to 3 feet (94 centimeters) deep, and cover about 4,000 square miles (10,300 square kilometers). Water comes from rainfall and overflows from Lake Okeechobee.

> ### THE FLORIDA EVERGLADES
> **Location:** Southern Florida
>
> **Area:** Approximately 1,506,539 acres (606,688 hectares)
>
> **Classification:** Freshwater marsh, freshwater swamp, and saltwater swamp

Plant life is diverse. Much of the Everglades is covered with sawgrass. Tropical plants, such as ferns, orchids, and mosses, grow in freshwater areas. Saltwater aquatic plants include lilies and bladderworts. Cypresses, mangroves, palms, live oaks, and pines grow on the hummocks.

Among the invertebrates, the Liguus tree snail can be found living on the hummocks, and Everglades reptiles include turtles, king snakes, water moccasins, and rattlesnakes. Alligators live in the freshwater areas while American crocodiles live in the salt water swamps.

Bird life in the Everglades attracts much tourist attention. Herons, egrets, spoonbills, ibises, eagles, and kites are a few of the majestic birds in the area. The Everglades kite, a tropical bird of prey, has been named after the region. Mammals include white tailed deer, cougars, bobcats, black bears, and otters.

Many attempts have been made to drain the Everglades so the land could be used for farming and development. Drainage projects began in the early 1900s. Consequently, about 50 percent of the wetlands have been altered to form water conservation areas. Because of agriculture and development, only about 20 percent of the original area has been preserved.

Draining of the area has also caused a loss of animal and plant life. Endangered species include the manatee, the round-tailed muskrat, and the mangrove fox squirrel. Early in the twentieth century, plume hunters nearly wiped out egrets and spoonbills for their feathers, which were used extensively as decorations on women's hats .

The introduction of non-native species of plants is also a threat. The Australian melaleuca tree and the Brazilian pepper plant have spread throughout the area, using up water supplies and choking native plants.

Agriculture is the leading economic activity in the region. In the area south of Lake Okeechobee, farmers raise sugarcane, fruit, and vegetables, which are shipped north. Sport fishing is popular in the area, as is boating, hunting, and camping.

The Seminole Indians, who were driven out of the Okefenokee Swamp in northern Florida by American troops during the nineteenth century, found a safe home in the Everglades. They planted corn and vegetables and gathered roots and nuts. Fishing and hunting were also sources of food. Throughout the twentieth century, the number of Seminoles living in the area has decreased. However, another tribe, the Miccosukee, still live in reservations near the center of the Everglades.

A growing concern for the ecology has encouraged protection of the Everglades and its ecosystem. It became a national park in 1947 and is the third largest and the wettest national park in the United States. The park includes 10,000 islands off the Florida coast along the Gulf of Mexico, and more than 1,000,000 people visit annually.

OKEFENOKEE SWAMP

The word "Okefenokee" (oh-kee-fen-OH-kee) is taken from a Timucuan Indian word and means "trembling earth." This refers to the small bushes and waterweeds that float on the swamp's open water and move when the water is disturbed. The swamp, which is 25 miles (40 kilometers) wide and 35 miles (56 kilometers) long, is a primitive, subtropical wilderness.

The Okefenokee lies about 50 miles (80 kilometers) inland from the coast of the Atlantic Ocean, but a sandy ridge prevents the swamp from draining directly into the ocean. Instead it is partially drained by the St. Mary's River, which flows into the Atlantic, and the Suwannee River, which flows into the Gulf of Mexico.

The Okefenokee was formed almost 250,000 years ago when the waters of the Atlantic Ocean covered the area. When the ocean receded, some salt water was trapped in depressions in the ground. Over thousands of years, plants grew and filled the wetland. As they decayed, they turned into peat, which today forms the swamp floor. The peat is as deep as 15 feet (4.6 meters) in places.

The Okefenokee also includes freshwater lakes, wet savannas (grasslands), cypress woods, hummocks, open grassy spaces, peat bog islands, and thick brush. Many channels form a maze through the area. About 50 inches (127 centimeters) of rain falls annually.

Rare orchids and lilies can be found here. The insectivorous pitcher plant, maiden cane, floating hearts, and golden club all add color. Freshwater regions support sphagnum moss, ferns, and rushes. Giant tupelo trees and bald cypress trees covered with Spanish moss add an eerie quality.

OKEFENOKEE SWAMP

Location: Southeastern Georgia and northern Florida

Area: 396,000 acres (158,400 hectares)

Classification: Freshwater swamp, bog, marsh

Insects, such as the mosquito and the yellow fly, are abundant, and 20 species of toads and frogs make the Okefenokee their home. At least 75 species of reptiles live here, including 5 kinds of poisonous snakes, and more than 10,000 alligators. Considered the "kings" of the swamp, the alligators are a major tourist attraction. About 40 species of fish live in the swamp, including bluegill, warmouth, golden shiners, and pumpkinseed sunfish. These fish have adapted to life in the acidic water. They are often dark with yellow undersides and they can see in low light.

The Okefenokee includes more than 200 species of birds, among which are the white ibis, the sandhill crane, the wood duck, the great blue heron, and the barred owl.

Bears, bobcats, deer, foxes, and raccoons live in the forested areas. Altogether, more than 40 species of mammals can be found in the swamp, including the endangered black bear.

The surrounding area around the swamp was settled by Europeans in the late 1700s. The Seminoles once lived in the interior, but they were driven out by American troops in 1838.

During the 1800s, developers tried to drain the swamp but found the project larger than anticipated. After the turn of the century, a railroad was built through the area and, as a result, many of the 500-year-old cypress trees were cut down. In 1937, U.S. president Franklin Roosevelt (1882–1945) protected about 460 square miles (1,196 square kilometers) of the Okefenokee, by designating it the Okefenokee National Wildlife Refuge. Despite these actions, the swamp continues to be threatened by developers, including those who want to mine titanium ore on its eastern rim. Mining requires the use of vast amounts of water, which would have to be pumped from the wetland.

Tourists can still enjoy the splendor and eerie wilderness of the Okefenokee, which offers walking trails and 107 miles (172 kilometers) of water trails that can be explored by canoe.

THE MEKONG DELTA

The Mekong Delta extends from Phnom Penh, the capital of Kampuchea province in Vietnam, to the coastline bordering the South China Sea. The Mekong River, the longest in Southeast Asia, divides at Phnom Penh into two branches. As the branches pass through the delta they form nine channels. The name Mekong means "Nine Dragons" and refers to these channels.

The delta forms a triangular area, including floodplain, freshwater, and saltwater swamps. Mangrove and melaleuca forests cover the remaining area.

Most of the delta is less than 16 feet (5 meters) above sea level. From May to October the area experiences heavy rains, ranging from 59 inches (150 centimeters) to 92 inches (235 centimeters). The average temperature is about 79°F (26°C), and the climate is very humid.

> ### THE MEKONG DELTA
> **Location:** Vietnam
> **Area:** 19,120 square miles (49,712 square kilometers)
> **Classification:** Freshwater swamp and saltwater swamp

The Mekong Delta supports 35 species of reptiles, including crocodiles and the endangered river terrapin, a type of turtle, and 260 species of fish, many which are sold commercially. Herons, egrets, ibises, and storks nest in the delta's forested areas. Mammals found there include otters and fishing cats.

The delta is rich in agricultural produce, fish, and waterfowl. At one time it yielded 26 tons (23.5 metric tons) of fish per square mile. About half of the delta is devoted to rice paddies, which yield about 6,500,000 tons (5,900,000 metric tons) of rice each year.

During the Vietnam War (1959–75), the Mekong Delta was seriously damaged when Agent Orange and other defoliants (chemicals that cause the leaves to fall off plants and trees) destroyed over 50 percent of the mangrove forests. The object was to remove any hiding places used by the enemy, but Agent Orange also caused skin diseases, cancer, and birth defects. One fifth of the country's farmland was destroyed by direct bombing and machines used to clear land. Also, the melaleuca forests were burned with Napalm, another chemical weapon, and almost destroyed. These chemicals also made much of the soil infertile.

An endangered bird, the eastern Sarus crane, disappeared during the war but has since returned. The giant ibis, the white-shouldered ibis, and the white-winged wood duck have not been seen here since 1980.

Hydroelectric dams, built for generating electricity, also have a negative effect on the ecosystem. The dams reduce flooding, which decreases the freshwater flow to wetlands. Spawning areas for fish are declining and mangrove forests will soon degrade without enough fresh water.

THE GREAT DISMAL SWAMP

The Great Dismal Swamp is a freshwater wetland called a pocosin, or upland swamp. Its soil contains few minerals and nutrients. Drainage is poor and the soil is saturated and acidic. In this environment, decomposition is slow and peat accumulates over time.

The fungus, armillaria, which grows on rotten wood in the swamp, gives off a luminescence (glow) called foxfire. Also at home here are broad-leaved shrubs such as hollies and bayberries, and the swamp is the only known habitat for the rare dwarf trillium and silky camillia. Bald cypress trees, which survive well in saturated soil, populate the area, some as old as 1,500 years. Other native trees include the black gum, the juniper, and the white ash.

> ## THE GREAT DISMAL SWAMP
> **Location:** North Carolina and Virginia
> **Area:** 106,716 acres (42,686 hectares)
> **Classification:** Peat bog and freshwater swamp

Seventy-three species of butterflies and more than 97 species of perching birds, including warblers, woodpeckers, and wood ducks, live in this area, as does the rare ivory-billed woodpecker. The Great Dismal Swamp is the only known habitat of the short-tailed shrew, a type of mouse. Black bears, bobcats, minks, and otters also live in the region.

Many famous Americans were in some way associated with the Great Dismal Swamp. In 1763, U.S. president George Washington (1732–1799) set up a company to drain the swamp for agricultural purposes and to use its lumber. The idea failed and the swamp survived. The American statesman, Patrick Henry (1736–1799), who participated in the American Revolution, owned land there. Also, the swamp was used by American author Harriet Beecher Stowe (1811–1892) as the setting for her antislavery novel *Uncle Tom's Cabin.*

KAKADU NATIONAL PARK

Kakadu National Park is the largest wetland in Australia and is marked by a great diversity in species and habitats. The area includes freshwater wetlands, mangrove swamps, and billabongs. The term billabong comes from an Australian Aboriginal (native) word meaning "dead river," and describes standing, often stagnant, water near river channels. Billabongs are usually found only in the parts of Australia where the climate is hot and dry and interrupted by occasional flooding.

Average temperatures are high all year and range from 91°F (33°C) in July to 108°F (42°C) in October. Rivers flooded by the January and February rains feed the freshwater areas. The coastal wetlands are fed by tides.

Kakadu also includes many acres of forests, sedges, and grasslands. In April and May, when the dry season begins, bush fires destroy old growth and encourage new plant life.

When the thousands of acres of grassland flood during the wet season, plant and animal life abounds. In all, there are more than 200 species of plants, including 22 species of mangrove trees. Water lilies, woolytbutt eucalyptus miniata, and spear grass, which grows to more than 6 feet (1.8 meters) tall, can all be found here. Paperbark trees dominate freshwater swamp areas.

> **KAKADU NATIONAL PARK**
>
> **Location:** East of Darwin in the Northern Territory of Australia
>
> **Area:** 7,700 square miles (20,000 square kilometers)
>
> **Classification:** Freshwater wetlands, mangrove swamps

Numerous turtles and both freshwater and saltwater crocodiles live in the shallows. Barramundi fish travel between the waterholes and estuaries, depending on the season. Goannas, a type of monitor lizard, also live here.

Millions of water birds live in the park. These include whistling ducks, magpie geese, radjal shelducks, and grey teals. Shorebirds migrate from as far away as the Arctic to winter in the mild climate of Kakadu.

The Australian native people, called Aborigines, once used the park's wetland plants and animals as a source of food. However, the disposal of waste from uranium mines on the edge of Kakadu is threatening the area. Other dangers include tourism and the introduction of non-native plants and animals. Para grass, an exotic grass being tested as cattle feed, is currently choking Kakadu's wetlands. The big-headed ant, a native of Africa, forms huge colonies that wipe out many native ants and other small creatures.

KUSHIRO MARSH

The Kushiro (koo-SHEE-roh) Marsh is one of the largest and most important natural wetlands in Japan. It is located in a floodplain and contains small freshwater lakes, reed beds, sedge marshes, and peat bogs. About 12,400 acres (5,000 hectares) have been designated as a national wildlife protection area and a national monument.

Annual rainfall averages about 44 inches (112 centimeters). The climate is on the cold side with winter temperatures sometimes falling below -4°F (-20°C). Spring and summer are also cool.

The floodplain areas support reeds and sedges. The shallow marshes are extremely important for the endangered red-crowned crane, because they provide a place for the birds to breed and winter. More than 100 species of birds use the area, including geese, ducks, and swans. Mammals found in the marsh include the raccoon dog and the red fox.

KUSHIRO MARSH

Location: The island of Eastern Hokkaido, Japan, along the Kushiro and Akan Rivers

Area: 112 square miles (290 square kilometers)

Classification: Marsh and peat bog

THE LLANOS

The Llanos (YAH-nos), a Spanish word for "plains," is a large grassland that covers parts of western Venezuela and northeastern Columbia. Within its boundaries is one of the largest wetlands in South America—a combination of rivers, lakes, marshes, swamps, forests, and ponds. It is bordered by the Andes Mountains on the north and west and the Orinoco and Apure Rivers to the east.

The Llanos is a temperate area, with year-round average temperatures of 75°F (24°C). Rain falls primarily between April and November, and the central plains get about 45 inches (110 centimeters) annually. Areas closer to the Andes Mountains receive about 180 inches (457 centimeters). The summer is very dry. No growth takes place during these months, and frequent fires drive animals to areas that are still wet. When the rains come in the winter, the growing season begins.

Although growth is limited by the long dry period, swamp grasses and sedges are common. Carpet grass is able to grow in drier areas. Oak and palm trees are found along the edges of the rivers.

THE LLANOS

Location: Venezuela and Columbia, South America

Area: 220,000 square miles (570,000 square kilometers)

Classification: Riparian wetland and swamp

The Llanos provides a habitat for many wading birds such as herons, storks, and ibises. Burrowing rodents are the only indigenous (native) mammals. The capybara, which is the world's largest rodent, lives in the area. It grows up to 4 feet (1.3 meters) long and weighs as much as 110 pounds (50 kilograms). Armadillos, deer, and anteaters find shelter in the forest and feed in the grasslands.

Sadly, too much hunting has negatively affected the wildlife, and the overgrazing of cattle has ruined much of the soil. Much small farming is also done in the area, which also impacts the environment.

OKAVANGO DELTA

The Okavango Delta is considered an oasis in the Kalahari Desert. As the Okavango River flows from Angola and into Botswana, it becomes choked with weeds and spreads out to create one of the world's largest inland wetlands.

Annual temperatures range from 21° to 105°F (-6° to 41°C). Average rainfall is about 18 inches (45 centimeters). In February and March, the river over-flows and refills the marshes and swamp areas, making them desirable habitats for plants and animals. Water can be as deep as 16 feet (about 5 meters).

Although rainfall can vary, parts of the delta are always wet from flooding. Water lilies and papyrus are abundant.

The delta is home to kingfishers, herons, and water birds such as ducks, geese, and ibises.

> **OKAVANGO DELTA**
>
> **Location:** Northwest Botswana, Africa
>
> **Area:** 6,175 square miles (16,000 square kilometers) in the dry season increasing to 8,500 square miles (22,000 square kilometers) in the wet season
>
> **Classification:** Freshwater marsh and swamp

When the waters recede, grazing animals arrive. These include antelope, springbok, and lechwe. The lechwe is a semi-aquatic antelope that feeds on grasses in shallow water. It can graze while standing in water up to 20 inches (5 centimeters) deep. Its elongated hooves help it move through the soft mud on the swamp floor.

The Okavango is the only source of permanent surface water for Botswana. This creates industrial and agricultural demands, such as for diamond mining. However, an area about 695 square miles (1,800 square kilometers) in size has been designated as the Moremi Wildlife Reserve, one of the few protected areas in Africa.

PRIPET MARSHES

The Pripet Marshes (also spelled Prepay or Pipit) form a very large wetland in Eastern Europe, covering the southern part of Belarus and northern Ukraine. The marshes are located in a forested basin of the Privet River, and the swamp is the largest in Europe.

Formed during the last Ice Age, which ended about 10,000 years ago, the marshes were filled with sand and gravel left behind as the glaciers melted. The many lakes in the area are in the process of turning into bogs. Sandy lowlands can be found among a maze of rivers, and pine forests and floodplains add to the diversity of the ecosystem. About one-third of the region is forested.

> **PRIPET MARSHES**
>
> **Location:** Southern Belarus and northern Ukraine
>
> **Area:** Approximately 104,000 square miles (270,000 square kilometers)
>
> **Classification:** Marsh, bog, and swamp

The climate is cool, with winter temperatures ranging from 18 to 25°F (-8 to -4°C). Warmer temperatures are found in and around the marshes. Annual precipitation is 22 to 26 inches (55 to 65 centimeters).

Trees growing here include pine, birch, alder, oak, aspen, white spruce, and hornbeam. The hornbeam is a type of birch that has smooth, gray bark and catkins (drooping scaly flowers with no petals).

Many types of birds live in the marsh, including orioles, grouse, woodpeckers, owls, blue tits, and ducks. Mammals include lynxes, wolves, foxes, wild boars, Asian deer, beavers, badgers, and weasels.

Much of the land has been cleared for lumber and agricultural use over the last several hundred years. Attempts to drain the marsh began in the mid-nineteenth century and are still underway. Crops grown in the area include rye, barley, wheat, flax, potatoes, and vegetables. Grasses used as cattle feed are also grown here.

THE PINE BARRENS

The Pine Barrens (or Pineland) are located on the outer coastal plain of Long Island and New Jersey, including parts of the Delaware River and the Jersey Shore. Much of the area is open forest broken by marshes, swamps, and bogs. The area was formed during the last Ice Age.

The northwest part of the Pine Barrens experiences relatively cold winters, with average January temperatures of less than 28°F (-2°C). The southern area is milder, with average winter temperatures above freezing. Summers are hot, with averages for July ranging from about 70°F (21°C) in the northwest to more than 76°F (24°C) in the southwest. Precipitation (rain, sleet, or snow) is evenly distributed throughout the seasons, averaging from 44 inches to more than 48 inches (112 to 122 centimeters) annually.

> **THE PINE BARRENS**
> **Location:** Long Island, New Jersey
> **Area:** Over 1,000,000 acres (400,000 hectares)
> **Classification:** Bog, swamp, and marsh

The Pine Barrens are home to about 100 endangered species of plants. Many rare species grow here, including the glade cress, great plains ladies-tresses, grooved yellow flax, and some insectivorous plants. Other plants include wild azaleas, purple cone flowers, Indian grasses, and little bluestems. Rhododendrons, honeysuckles, mountain laurels, wintergreens, and cardinal flowers bring color to the area.

The area is dominated by oak and pine trees, which thrive on the well-drained sites. White cedars grow in the poorly drained bogs. Other trees include buckthorns, dogwoods, sugar maples, hemlocks, birches, ashes, and sweet gums.

The endangered Pine Barrens tree frog makes its home in the park, while bears and wildcats can still be found in some of the woodland areas. Deer, opossums, and raccoons are also common.

Commercially, the area is valued for its production of blueberries and cranberries and for tourism.

Designated by Congress in 1978 as the country's first National Reserve, the Pine Barrens' natural and cultural resources are now protected.

POLAR BEAR PROVINCIAL PARK

Polar Bear Provincial Park is on the northwest coast of James Bay and the southern coast of Hudson Bay in Ontario, Canada. A true wilderness, it is accessible only by plane or boat. The dominant type of wetland in the park is peatland. Inland areas of the park contain swamps and marshes, while coastal areas contain saltwater marshes.

Many of the wetlands have formed in kettles (depressions in the ground left behind by retreating glaciers). Rain, melted snow, and overflowing rivers contribute to the water supply. As kettle lakes fill in with vegetation, marsh are formed, and coastal marshes have developed with the movement of the tides. The flat-topped ridges that run along the beach, parallel to the coast, help prevent drainage back into the sea.

Because the park lies along the southern edge of the Arctic region, temperatures are cold.

> **POLAR BEAR PROVINCIAL PARK**
> **Location:** Ontario, Canada
> **Area:** 9,300 square miles (24,087 square kilometers)
> **Classification:** Peatland, swamp, freshwater marsh, and saltwater marsh

Freezing weather prevails for six to eight months. Summers are short with the average temperature between 60° to 70°F (16° to 25°C). Precipitation is usually less than 21 inches (55 centimeters).

Vegetation varies, depending on altitude and wetness. In the higher, cooler regions of the park, plant life includes sedges, goose grasses, and flowering saxifrages. Cotton grasses, sedges, and birches grow in the lower areas. In drier places, blueberries, crowberries, and louseworts are found. Salt-tolerant plants grow in the saltwater marshes, including aquatic grasses, cotton grasses, lungworts, and lyme grasses. Caribou lichens and reindeer mosses grow in wetlands that are closer to forested areas.

Many waterfowl, such as the lesser snow goose, use the park in spring and fall as they migrate to their Arctic breeding grounds. Thousands of mallard ducks pass through in the fall, and the western and southwestern parts of the park form a migration path for shorebirds, such as the ruddy turnstone, black-bellied plover, and several species of sandpipers. Canadian geese also nest in the park.

Polar bears spend their summers in the park while breeding. Caribou and moose also roam the area, moving north in the warmer weather. Beavers, muskrats, otters, foxes, and wolves are just a few of the many other mammals that make the park their year-round home.

The Cree Indians live along the coastal areas and use the park for hunting, fishing, and trapping. The Cree own two hunting and fishing camps, where guests can fish and hunt waterfowl, grouse, and snipe. No other non-native hunting or fishing is permitted.

Since 1970, the area has been protected from development, and commercial or industrial use of natural resources is prohibited. Most of the park is designated as wilderness zones, nature reserves, or historical zones, where wildlife is protected.

USUMACINTA DELTA

The Usumacinta (oo-soo-mah-SEEN-tah) Delta is the most extensive wetland on the Gulf Coast of Mexico. It contains freshwater lagoons, swamps, marshes, and mangrove swamps. Its waters are rich in nutrients and support a major coastal fishery.

> **USUMACINTA DELTA**
>
> **Location:** Gulf Coast of Mexico
>
> **Area:** 2.47 million acres (998,000 hectares)
>
> **Classification:** Swamp, marsh, and mangrove swamp

Many waterbirds breed and winter in the area. These include herons, egrets, storks, ibises, and spoonbills. Shorebirds that winter in the delta are ducks and coots.

A few manatees from the West Indies also can be found here, as well as the endangered Morelet's crocodile.

Threats to the delta have come from drainage for agriculture and oil spills from a nearby oil field. Also, mangrove trees are being cut for timber.

FLOW COUNTRY

> **FLOW COUNTRY**
>
> **Location:** Northeast corner of Scotland
>
> **Area:** 902,681 square acres (365,310 hectares)
>
> **Classification:** Blanket bog

The marshy area of Caithness and Sutherland countries is known as the Flow Country. The name "flow" comes from an old Norse word meaning "marshy ground." The Flow Country is one of the largest blanket bogs in the world, and it is still growing. Some of the peat found here is more than 7,000 years old.

Typical plants that grow in the Flow Country include sphagnum moss, heather, purple moorgrass, sedges, and rushes. The insect-eating plant, sundew, also inhabits the bog.

Many bird species are supported by the bog. Seventy percent of the greenshanks in Great Britain and the entire population of black-throated divers live there.

Because the peat that grows in a blanket bog is very valuable as fuel for home heating and industrial use, peat mining is endangering this ancient ecosystem. In the 1980s, the Scottish government and several private companies began planting pine and fir trees. The trees support the forestry industry, but will eventually take over the wetland.

FOR MORE INFORMATION

BOOKS

Fowler, Allan. *Life in a Wetland.* Danbury, CT: Children's Press, 1999.

Gilman, Kevin. *Hydrology and Wetland Conservation.* New York: John Wiley & Sons, Inc., 1994.

Grisecke, Ernestine. *Wetland Plants.* Portsmouth, NH: Heinemann Library, 1999.

Taylor, David. *Endangered Wetland Animals.* New York: Crabtree Publishing, Co., 1992.

Thompson, Wynne. *Journey to a Wetland.* Huntington, WV: Aegina Press, Inc., 1997.

ORGANIZATIONS

Environmental Protection Agency
  Office of Wetlands, Oceans and Watersheds
  Wetlands Division (4502F)
  401 M Street SW
  Washington, DC 20460
  Phone: 202-260-2090
  Internet: http://www.epa.gov

National Wetlands Conservation Project
  The Nature Conservancy
  1800 N Kent Street, Suite 800
  Arlington, VA 22209
  Phone: 703-841-5300; Fax: 703-841-1283
  Internet: http://www.tnc.org

WEBSITES

Note: Website addresses are frequently subject to change.

Thurston High School, Biomes: http:// ths.sps.lane.edu/biomes/index1.html

University of California at Berkeley: http://www.ucmp.berkeley.edu/glossary/gloss5/biome/index.html

BIBLIOGRAPHY

"America's Wetlands." Environmental Protection Agency. http://www.epa.gov/OWOW/wetlands/vital/what.html

"Basic Soils." University of Minnesota. http://www.soils.agi.umn/edu/academics/classes/soil3125/index.html

Borgioli, A., and G. Cappelli. *The Living Swamp.* London: Orbis, 1979.

Capstick, Peter Hathaway. *Death in the Long Grass.* New York: St. Martin's Press, 1977.

Duffy, Trent. *The Vanishing Wetlands.* New York: Franklin Watts, 1994.

Dugan, Patrick, ed. *Wetlands In Danger: A World Conservation Atlas.* New York: Oxford University Press, 1993.

Everard, Barbara, and Brian D. Morley. *Wild Flowers of the World.* New York: Avenal Books, 1970.

Everglades: Encyclopedia of the South. New York: Facts on File, 1985.

Finlayson, Max, and Michael Moser. *Wetland.* New York: Facts on File, 1991.

Franck, Irene, and David Brownstone. *The Green Encyclopedia.* New York: Prentice Hall, 1992.

Ganeri, Anita. *Ecology Watch: Rivers, Ponds and Lakes.* New York: Dillon Press, 1991.

George, Jean Craighead. *Everglades Wildguide.* Washington, DC: National Park Service, U.S. Dept. of the Interior, 1988.

Hirschi, Ron. *Save Our Wetlands.* New York: Delacorte Press, 1995.

Levathes, Louise E. "Mysteries of the Bog." *National Geographic.* Vol 171, No. 3, March 1987, pp. 397–420.

Leyder, Francois. "Okefenokee, The Magical Swamp." in *National Geographic.* Vol. 145, No. 2, Feb. 1974, pp. 169–75.

McCormick, Anita Louise. *Vanishing Wetlands.* San Diego: Lucent Books, 1995.

McLeish, Ewan. *Habitats: Wetlands.* New York: Thomson Learning, 1996.

Niering, William A. *Wetlands.* New York: Alfred A. Knopf, 1924.

"On Peatlands and Peat." International Peat Society. http://www.peatsociety.fi.

Ownby, Miriam. *Explore the Everglades.* Kissimmee, FL: Teakwood Press, 1988.

Parker, Steve. *Eyewitness Books: Pond and River.* New York: Alfred A. Knopf, 1988.

"Pripet Marshes" Britannica Online. http://www.eb.com:180/cgi-bin/g?DocF=micro/481/91.html

Rezendes, Paul, and Paulette Roy. *Wetlands: The Web of Life.* Vermont: Verve Editions, 1996.

Rotter, Charles. *Wetlands.* Mankato, MN: Creative Education, 1994.

Sayre, April Pulley. *Wetland. Exploring Earth's Biomes.* New York: Twenty-First Century Books, 1996.

Sheldrake, Rupert. *A New Science of Life: The Hypothesis of Formative Causation.* Los Angeles: J. P. Tarcher, Inc., 1981.

Sheldrake, Rupert. *Seven Experiments That Could Change the World.* New York: Riverhead Books, 1995.

Simon, Noel. *Nature in Danger.* New York: Oxford University Press, 1995.

"Soil." Microsoft Encarta @ 97 Encyclopedia. © 1993-1996 Microsoft Corporation.

Staub, Frank. *America's Wetlands*. Minneapolis, MN: Carolrhoda Books, Inc., 1995.

Swink, Floyd, and Gerould Wilhelm. *Plants of the Chicago Region*. Lisle, IL: The Morton Arboretum, 1994.

Wetlands: Meeting the President's Challenge. U.S. Dept. of the Interior, U.S. Fish and Wildlife Service, 1990.

"Wetlands." World Book Encyclopedia. Chicago: Scott Fetzer, 1996.

BIBLIOGRAPHY

BOOKS

Aldis, Rodney. *Ecology Watch: Polar Lands*. New York: Dillon Press, 1992.

"America's Wetlands." Environmental Protection Agency, http://www.epa.gov/OWOW/wetlands/vital/what.html

Amos, William H. *The Life of the Seashore*. New York: McGraw-Hill Book Company, 1966.

"Arctic National Wildlife Refuge." U. S. Fish and Wildlife Services, http://www.r7.fws.gov/nwr/arctic/arctic.html

Arritt, Susan. *The Living Earth Book of Deserts*. New York: Reader's Digest, 1993.

Ayensu, Edward S., ed. *Jungles*. New York: Crown Publishers, Inc., 1980.

Baines, John D. *Protecting the Oceans*. Conserving Our World. Milwaukee, WI: Raintree Steck-Vaughn, 1990.

Barber, Nicola. *Rivers, Ponds, and Lakes*. London: Evans Brothers Limited, 1996.

Barnhart, Diana, and Vicki Leon. *Tidepools*. Parsippany, NJ: Silver Burdett Press, 1995.

"Basic Soils." University of Minnesota: http://www.soils.agi.umn.edu/academics/classes/soil3125/index.html

"Beach and Coast." *Grolier Multimedia Encyclopedia*. Danbury, CT: Grolier, Inc., 1995.

Beani, Laura, and Francesco Dessi. *The African Savanna*. Milwaukee, WI: Raintree Steck-Vaughn Publishers, 1989.

Borgioli, A., and G. Cappelli. *The Living Swamp*. London: Orbis, 1979.

Brown, Lauren. *Grasslands*. New York: Alfred A. Knopf, 1985.

Burnie, David. *Tree*. New York: Alfred A. Knopf, 1988.

Burton, John A. *Jungles and Rainforests*. The Changing World. San Diego, CA: Thunder Bay Press, 1996.

Burton, Robert, ed. *Nature's Last Strongholds*. New York: Oxford University Press, 1991.

Campbell, Andrew. *Seashore Life*. London: Newnes Books, 1983.

Capstick, Peter Hathaway. *Death in the Long Grass*. New York: St. Martin's Press, 1977.

Catchpoh, Clive. *The Living World: Mountains*. New York: Dial Books for Young Readers, 1984.

Caras, Roger. *The Forest: A Dramatic Portrait of Life in the American Wild*. New York: Holt, Rinehart, Winston, 1979.

Carroll, Michael W. *Oceans and Rivers*. Colorado Springs, CO: Chariot Victor Publishing, 1999.

BIBLIOGRAPHY

Carwardine, Mark. *Whales, Dolphins, and Porpoises.* See and Explore Library. New York: Dorling Kindersley, 1992.

Cerullo, Mary M. *Coral Reef: A City That Never Sleeps.* New York: Cobblehill Books, 1996.

"Chaparral." Britannica Online.http://www.eb.com:180/cgi-bin/g? DocF= micro/116/70.html

Christiansen, M. Skytte. *Grasses, Sedges, and Rushes.* Poole, Great Britain: Blandford Press, 1979.

Claybourne, Anna. *Mountain Wildlife.* Saffron Hill, London: Usborne Publishing, Ltd., 1994.

Cloudsley-Thompson, John. *The Desert.* New York: G.P. Putnam's Sons, 1977.

Cole, Wendy. "Naked City: How an Alien Ate the Shade." *Time.* (February 15, 1999): 6.

Cricher, John C. *Field Guide to Eastern Forests.* Peterson Field Guides. Boston: Houghton Mifflin, 1988.

Cumming, David. *Coasts.* Habitats. Austin, TX: Raintree Steck-Vaughn, 1997.

Curry-Lindahl, Kai. *Wildlife of the Prairies and Plains.* New York: Harry N. Abrams, Inc., 1981.

Davis, Stephen, ed. *Encyclopedia of Animals.* New York: St. Martin's Press, 1974.

"Desert." *Encyclopaedia Britannica.* Chicago: Encyclopaedia Britannica, Inc., 1993.

"Desert." *Grolier Multimedia Encyclopedia.* New York: Grolier, Inc., 1995.

Dixon, Dougal. *The Changing Earth.* New York: Thomson Learning, 1993.

Dixon, Dougal. *Forests.* New York: Franklin Watts, 1984.

Dudley, William. *Endangered Oceans.* Opposing Viewpoints Series. San Diego, CA: Greenhaven Press, Inc., 1999.

Duffy, Eric. *The Forest World: The Ecology of the Temperate Woodlands.* New York: A&W Publishers, Inc., 1980.

Duffy, Trent. *The Vanishing Wetlands.* New York: Franklin Watts, 1994.

Dugan, Patrick, ed. *Wetlands In Danger: A World Conservation Atlas.* New York: Oxford University Press, 1993.

Engel, Leonard. *The Sea.* Life Nature Library. New York: Time-Life Books, 1969.

Everard, Barbara, and Brian D. Morley. *Wild Flowers of the World.* New York: Avenal Books, 1970.

Everglades: Encyclopedia of the South. New York: Facts on File, 1985.

Feltwell, John. *Seashore.* New York: DK Publishing, Inc., 1997.

Finlayson, Max, and Michael Moser. *Wetland.* New York: Facts on File, 1991.

Fisher, Ron. *Heartland of a Continent: America's Plains and Prairies.* Washington, DC: National Geographic Society, 1991.

Fitzharris, Tim. *Forests.* Canada: Stoddart Publishing Co., 1991.

Flegg, Jim. *Deserts: A Miracle of Life.* London: Blandford Press, 1993.

"Forest." *Colliers Encyclopedia.* CD-ROM, P. F. Collier, 1996.

"Forestry and Wood Production; Forestry: Purposes and Techniques of Forest Management; Fire Prevention and Control." Britannica Online. http://www.eb.com:180/cgi-bin/g? DocF=macro/5002/41/19.html

"Forests." Britannica Online. http://www.eb.com:180/cgi-bin/g?DocF= macro/5002/41/19.html

"Forests and Forestry." *Grolier Multimedia Encyclopedia.* New York: Grolier, Inc., 1995.

Forman, Michael. *Arctic Tundra.* New York: Children's Press, 1997.

Fornasari, Lorenzo, and Renato Massa. *The Arctic*. Milwaukee, WI: Raintree Steck-Vaughn Publishers, 1989.

Fowler, Allan. *Arctic Tundra*. Danbury, CT: Children's Press, 1997.

Fowler, Allan. *Life in a Wetland*. Danbury, CT: Children's Press, 1999.

Franck, Irene, and David Brownstone. *The Green Encyclopedia*. New York: Prentice Hall, 1992.

Ganeri, Anita. *Ecology Watch: Rivers, Ponds and Lakes*. New York: Dillon Press, 1991.

Ganeri, Anita. *Habitats: Forests*. Austin, Texas: Raintree Steck-Vaughn, 1997.

Ganeri, Anita. *Ponds and Pond Life*. New York: Franklin Watts, 1993.

George, Jean Craighead. *Everglades Wildguide*. Washington, DC: National Park Service, U.S. Dept. of the Interior, 1988.

George, Jean Craighead. *One Day in the Alpine Tundra*. New York: Thomas Y. Crowel, 1984.

George, Michael. *Deserts*. Mankato, Minnesota Creative Education, Inc., 1992.

Gilman, Kevin. *Hydrology and Wetland Conservation*. New York: John Wiley & Sons, Inc., 1994.

Gleseck, Ernestine. *Pond Plants*. Portsmouth, NH: Heinemann Library, 1999.

"Glow-in-the-Dark Shark Has Killer Smudge." *Science News* 154 (August 1, 1998): 70.

"Grass." *WorldBook Encyclopedia*. Chicago: World Book Incorporated, Scott Fetzer Co., 1999.

Greenaway, Theresa, Christiane Gunzi, and Barbara Taylor *Forest*. New York: DK Publishing, 1994.

Greenaway, Theresa. *Jungle*. Eyewitness Books. New York: Alfred A. Knopf, 1994.

Grisecke, Ernestine. *Wetland Plants*. Portsmouth, NH: Heinemann Library, 1999.

Gutnick, Martin J., and Natalie Browne-Gutnik. *Great Barrier Reef*. Wonders of the World. Austin, TX: Raintree Steck-Vaughn Publishers, 1995.

Grzimek, Bernhard. *Grizmek's Animal Encyclopedia: Fishes II and Amphibians*. New York: Van Nostrand Reinhold, 1972.

Grzimek, Bernhard. *Grizmek's Animal Encyclopedia: Reptiles*. New York: Van Nostrand Reinhold, 1972.

Hargreaves, Pat, ed. *The Indian Ocean*. Seas and Oceans. Morristown, NJ: Silver Burdett Company, 1981.

Hargreaves, Pat, ed. *The Red Sea and Persian Gulf*. Seas and Oceans. Morristown, NJ: Silver Burdett Company, 1981.

Hecht, Jeff. *Vanishing Life*. New York: Charles Scribner & Sons, 1993.

Hirschi, Ron. *Save Our Forests*. New York: Delacorte Press, 1993.

Hirschi, Ron. *Save Our Prairies and Grasslands*. New York: Delacorte Press, 1994.

Hirschi, Ron. *Save Our Wetlands*. New York: Delacorte Press, 1995.

Hiscock, Bruce. *Tundra: The Arctic Land*. New York: Atheneum, 1986.

Holing, Dwight. *Coral Reefs*. Parsippany, NJ: Silver Burdett Press, 1995.

The Hospitable Desert. Hauppauge, NY: Barron's Educational Series, Inc., 1999.

Inseth, Zachary. *The Tundra*. Chanhassen, MN: The Child's World, Inc., 1998.

International Book of the Forest. New York: Simon and Schuster, 1981.

Kaplan, Elizabeth. *Biomes of the World: Taiga*. New York: Benchmark Books, 1996.

Kaplan, Elizabeth. *Biomes of the World: Temperate Forest*. New York: Benchmark Books, 1996.

BIBLIOGRAPHY

Kaplan, Elizabeth. *Biomes of the World: Tundra.* New York: Marshall Cavendish, 1996.

Kaplan, Eugene H. *Southeastern and Caribbean Seashores.* NY: Houghton Mifflin, Company, 1999.

Khanduri, Kamini. *Polar Wildlife.* London: Usborne Publishing, Ltd., 1992.

Knapp, Brian. *Lake.* Land Shapes. Danbury, CT: Grolier Educational Corporation, 1993.

Knapp, Brian. *River.* Land Shapes. Danbury, CT: Grolier Educational Corporation, 1992.

Knapp, Brian. *What Do We Know About Grasslands?* New York: Peter Bedrick Books, 1992.

"Kola Peninsula." Britannica Online. http://www.eb.com:180/cgi-bin/g?DocF=micro/326/3.html

"Lake." *Encyclopaedia Britannica.* Chicago: Encyclopaedia Britannica, Inc., 1993.

"Lake." *Grolier Multimedia Encyclopedia.* Danbury, CT: Grolier, Inc., 1995.

"Lakes." *Britannica Online.* http://www.eb.com.180/cgi-bin/g?DocF=macro/5003/58.html

Lambert, David. *Our World: Grasslands.* Parsippany, NJ: Silver Burdett Press, 1988.

Lambert, David. *People of the Desert.* Orlando, FL: Raintree Steck-Vaughn Publishers, 1999.

Lambert, David. *Seas and Oceans.* New View. Milwaukee, WI: Raintree Steck-Vaughn Company, 1994.

Leopold, A. Starker. *The Desert.* New York: Time Incorporated, 1962.

Levathes, Louise E. "Mysteries of the Bog." *National Geographic.* Vol 171, No. 3, March 1987, pp. 397–420.

Levete, Sarah. *Rivers and Lakes.* Brookfield, CT: Millbrook Press, Inc., 1999.

Leyder, Francois. "Okefenokee, The Magical Swamp." *National Geographic.* Vol. 145, No. 2, Feb. 1974, pp. 169–75.

Lourie, Peter. *Hudson River: An Adventure from the Mountains to the Sea.* Honesdale, PA: Boyds Mills Press, 1992.

Lovelock, James. *The Ages of Gaia: A Biography of Our Living Earth.* New York: Bantam Books, 1990.

Lye, Keith. *Our World: Mountains.* Morristown, NJ: Silver Burdett, 1987.

Macquitty, Miranda. *Desert.* New York: Alfred A. Knopf, 1994.

Macquitty, Miranda. *Ocean.* Eyewitness Books. New York: Alfred A. Knopf, 1995.

Madson, John. *Tallgrass Prairies.* Helena, MT: Falcon Press Publishing, 1993.

Markle, Sandra. *Pioneering Ocean Depths.* New York: Atheneum, 1995.

Massa, Renato. *Along the Coasts.* Orlando, FL: Raintree Steck-Vaughn Publishers, 1998.

Massa, Renato. *India.* World Nature Encyclopedia. Milwaukee, WI: Raintree Steck-Vaughn Publishers, 1989.

McCormick, Anita Louise. *Vanishing Wetlands.* San Diego: Lucent Books, 1995.

McCormick, Jack. *The Life of the Forest.* New York: McGraw-Hill, 1966.

McLeish, Ewan. *Habitats: Oceans and Seas.* Austin, TX: Raintree Steck-Vaughn Company, 1997.

McLeish, Ewan. *Habitats: Wetlands.* New York: Thomson Learning, 1996.

Moore, David M., ed. *The Marshall Cavendish Illustrated Encyclopedia of Plants and Earth Sciences.* Vol. 7. New York: Marshall Cavendish, 1990.

Morgan, Nina. *The North Sea and the Baltic Sea.* Seas and Oceans. Austin, TX: Raintree Steck-Vaughn Company, 1997.

Morgan, Sally. *Ecology and Environment.* New York: Oxford University Press, 1995.

Nadel, Corinne J., and Rose Blue. *Black Sea.* Wonders of the World. Austin,

TX: Raintree Steck-Vaughn Company, 1995.

Nardi, James B. *Once Upon a Tree: Life from Treetop to Root Tips.* Iowa City, IA: Iowa State University Press, 1993.

National Park Service, Katmai National Park and Preserve. http://www.nps.gov/katm/

Niering, William A. *Wetlands.* New York: Alfred A. Knopf, 1924.

"Ocean." *Encyclopedia Britannica.* Chicago: Encyclopedia Britannica, Inc., 1993.

"Ocean and Sea." *Grolier Multimedia Encyclopedia.* Danbury, CT: Grolier, Inc., 1995.

The Ocean. Scientific American, Inc., San Francisco: W. H. Freeman and Company, 1969.

Oceans. The Illustrated Library of the Earth. Emmaus, PA: Rodale Press, Inc., 1993.

Ocean World of Jacques Cousteau. Guide to the Sea. New York: World Publishing, 1974.

"On Peatlands and Peat." International Peat Society, http://www.peatsociety.fi

Oram, Raymond F. *Biology: Living Systems.* Westerville, OH: Glencoe/McGraw-Hill, 1994.

Owen, Andy, and Miranda Ashwell. *Lakes.* Portsmouth, NH: Heinemann Library, 1998.

Ownby, Miriam. *Explore the Everglades.* Kissimmee, FL: Teakwood Press, 1988.

Parker, Steve. *Eyewitness Books: Pond and River.* New York: Alfred A. Knopf, 1988.

Page, Jake. *Planet Earth: Forest.* Alexandria, VA: Time-Life Books, 1983.

Pernetta, John. *Atlas of the Oceans.* New York: Rand McNally, 1994.

Pipes, Rose. *Grasslands.* Orlando, FL: Raintree Steck-Vaughn Publishers, 1998.

Pipes, Rose. *Tundra and Cold Deserts.* Orlando, FL: Raintree Steck-Vaughn Publishers, 1999.

Plants of the Desert. Broomall, PA: Chelsea House Publishers, 1997.

Pringle, Laurence. *Fire in the Forest: A Cycle of Growth and Renewal.* New York: Atheneum Books, 1995.

Pringle, Laurence. *Rivers and Lakes.* Planet Earth. Alexandria, VA: Time-Life Books, 1985.

"Pripet Marshes" Britannica Online. http://www.eb.com:180/cgi-bin/g?DocF=micro/481/91.html

Pritchett, Robert. *River Life.* Secaucus, NJ: Chartwell Books, Inc., 1979.

Quinn, John R. *Wildlife Survival.* New York: Tab Books, 1994.

Radley, Gail, and Jean Sherlock. *Grasslands and Deserts.* Minneapolis, MN: The Lerner Publishing Group, 1998.

Renault, Mary. *The Nature of Alexander.* New York: Pantheon Books, 1975.

Rezendes, Paul, and Paulette Roy. *Wetlands: The Web of Life.* Vermont: Verve Editions, 1996.

Ricciuti, Edward. *Biomes of the World: Chaparral.* New York: Benchmark Books, 1996.

Ricciuti, Edward. *Biomes of the World: Grassland.* New York: Benchmark Books, 1996.

Ricciuti, Edward R. *Biomes of the World: Ocean.* New York: Benchmark Books, 1996.

"River and Stream." *Grolier Multimedia Encyclopedia.* Danbury, CT: Grolier, Inc., 1995.

"River." *Encyclopaedia Britannica.* Chicago: Encyclopaedia Britannica, Inc., 1993.

Rootes, David. *The Arctic.* Minneapolis: MN: Lerner, 1996.

Rosenblatt, Roger. "Call of the Sea." *Time* (October 5, 1998): 58-71.

Rotter, Charles. *The Prairie.* Mankato, MN: Creative Education, 1994.

Rotter, Charles. *Wetlands.* Mankato, MN: Creative Education, 1994.

BIBLIOGRAPHY

Rowland-Entwistle, Theodore. *Jungles and Rainforests.* Our World. Morristown, NJ: Silver Burdett Press, 1987.

Sage, Bryan. *The Arctic and Its Wildlife.* New York: Facts on File, 1986.

Sanderson, Ivan. *Book of Great Jungles.* New York: Julian Messner, 1965.

Savage, Stephen. *Animals of the Desert.* Orlando, FL: Raintree Steck-Vaughn Publishers, 1996.

Savage, Stephen. *Animals of the Grasslands.* Orlando, FL: Raintree Steck-Vaughn Publishers, 1997.

"Saving the Salmon." *Time* (March 29, 1999): 60-61.

Sayre, April Pulley. *Desert.* Exploring Earth's Biomes. New York: Twenty-First Century Books, 1994.

Sayre, April Pulley. *Grassland.* New York: Twenty-First Century Books, 1994.

Sayre, April Pulley. *Lake and Pond.* New York: Twenty-First Century Books, 1996.

Sayre, April Pulley. *River and Stream.* New York: Twenty-First Century Books, 1995.

Sayre, April Pulley. *Seashore.* New York: Twenty-First Century Books, 1996.

Sayre, April Pulley. *Taiga.* New York: Twenty-First Century Books, 1994.

Sayre, April Pulley. *Temperate Deciduous Forest.* New York: Twenty-First Century Books, 1994.

Sayre, April Pulley. *Tundra.* New York: Twenty-First Century Books, 1994.

Sayre, April Pulley. *Wetland.* New York: Twenty-First Century Books, 1996.

Schoonmaker, Peter S. *The Living Forest.* New York: Enslow, 1990.

Schwartz, David M. *At the Pond.* Milwaukee, WI: Gareth Stevens, Inc., 1999.

Scott, Michael. *The Young Oxford Book of Ecology.* New York: Oxford University Press, 1995.

"Serengeti National Park" Britannica Online. http://www.eb.com:180/cgi-bin/g?DocF=micro/537/95.html

Sheldrake, Rupert. *A New Science of Life: The Hypothesis of Formative Causation.* Los Angeles: J. P. Tarcher, Inc., 1981.

Sheldrake, Rupert. *Seven Experiments That Could Change the World.* New York: Riverhead Books, 1995.

Shepherd, Donna. *Tundra.* Danbury, CT: Franklin Watts, 1997.

Silver, Donald M. *One Small Square: Arctic Tundra.* New York: W. H. Freeman and Company, 1994.

Silver, Donald M. *One Small Square: Pond.* New York: W. H. Freeman and Company, 1994.

Silver, Donald M. *One Small Square: Woods.* New York: McGraw-Hill, 1995.

Simon, Noel. *Nature in Danger.* New York: Oxford University Press, 1995.

Siy, Alexandra. *Arctic National Wildlife Refuge.* New York: Dillon Press, 1991.

Siy, Alexandra. *The Great Astrolabe Reef.* New York: Dillon Press, 1992.

Siy, Alexandra. *Native Grasslands.* New York: Dillon Press, 1991.

Slone, Lynn. *Ecozones: Arctic Tundra.* Viro Beach, FL: Rourke Enterprises, 1989.

"Soil" Microsoft Encarta 97 Encyclopedia. © 1993-1996 Microsoft Corporation.

Staub, Frank. *America's Prairies.* Minneapolis, MN: Carolrhoda Books, Inc., 1994.

Staub, Frank. *America's Wetlands.* Minneapolis, MN: Carolrhoda Books, Inc., 1995.

Staub, Frank. *Yellowstone's Cycle of Fire.* Minneapolis, MN: Carolrhoda Books, 1993.

Steele, Phillip. *Geography Detective: Tundra.* Minneapolis, MN: Carolrhoda Books, Inc., 1996.

Steele, Phillip. *Grasslands*. Minneapolis, MN: The Lerner Publishing Group, 1997.

Stidworthy, John. *Ponds and Streams*. Nature Club. Mahwah, NJ: Troll Associates, 1990.

Stille, Darlene R. *Grasslands*. Danbury, CT: Children's Press, 1999.

Storer, John. *The Web of Life*. New York: New American Library, 1953.

Sutton, Ann, and Myron Sutton. *Wildlife of the Forests*. New York: Harry N. Abrams, Inc., 1979.

Swink, Floyd, and Gerould Wilhelm. *Plants of the Chicago Region*. Lisle, IL: The Morton Arboretum, 1994.

Taylor, David. *Endangered Grassland Animals*. New York: Crabtree Publishing, 1992.

Taylor, David. *Endangered Wetland Animals*. New York: Crabtree Publishing, Co., 1992.

"Taymyr." Britannica Online. http://www.eb.com:180/cgi-Bin/g?DocF=micro/584/53.html

Thompson, Gerald, and Jennifer Coldrey. *The Pond*. Cambridge, MA: MIT Press, 1984.

Thompson, Wynne. *Journey to a Wetland*. Huntington, WV: Aegina Press, Inc., 1997.

"Thunderstorms and Lightning: The Underrated Killer." U.S. Department of Commerce. National Oceanic and Atmospheric Administration. National Weather Service, 1994.

"Tornado." Britannica Online. http://www.eb.com:180/cgi-bin/g?DocF=micro/599/23.html

"Tree." *Encyclopaedia Britannica*. Chicago: Encyclopaedia Britannica, Inc., 1993.

University of Wisconsin. "Stevens Point." http://www.uwsp.edu/acaddept/geog/faculty/ritter/geog101/biomes_toc.html

Wadsworth, Ginger. *Tundra Discoveries*. Walertown, MA: Charlesbridge Publishing, Inc., 1999.

Warburton, Lois. *Rainforests*. Overview Series. San Diego, CA: Lucent Books, Inc., 1990.

Waterlow, Julia. *Habitats: Deserts*. New York: Thomson Learning, 1996.

Waterflow, Julia. *Habitats: Grasslands*. Austin, TX: Raintree Steck-Vaughn Publishers, 1997.

Watson, Lyall. *Dark Nature*. New York: HarperCollins Publishers, 1995.

Wells, Susan. *The Illustrated World of Oceans*. New York: Simon and Schuster, 1991.

Wetlands: Meeting the President's Challenge. U.S. Dept. of the Interior, U.S. Fish and Wildlife Service, 1990.

"Wetlands." World Book Encyclopedia. Chicago: Scott Fetzer, 1996.

Whitfield, Philip, ed. *Atlas of Earth Mysteries*. Chicago: Rand McNally, 1990.

"Wind Cave National Park." http://www.nps.gov/wica/

Yates, Steve. *Adopting a Stream: A Northwest Handbook*. Seattle, WA: University of Washington Press, 1989.

Zich, Arthur. "Before the Flood: China's Three Gorges." *National Geographic*. (September, 1997): 2-33.

ORGANIZATIONS

African Wildlife Foundation
1400 16th Street NW, Suite 120
Washington, DC 20036
Phone: 202-939-3333
Fax: 202-939-3332
Internet: http://www.awf.org

American Cetacean Society
PO Box 1391
San Pedro, CA 90731
Phone: 310-548-6279
Fax: 310-548-6950
Internet: http://www.acsonline.org

American Littoral Society
Sandy Hook

Highlands, NJ 07732
Phone: 732-291-0055

American Oceans Campaign
725 Arizona Avenue, Suite. 102
Santa Monica, CA 90401
Phone: 800-8-OCEAN-0
Internet: http://www.americanoceans.
com

American Rivers
1025 Vermont Avenue NW, Suite 720
Washington, DC 20005
Phone: 800-296-6900
Fax: 202-347-9240
Internet: http://www.amrivers.org

Astrolabe, Inc.
4812 V. Street, NW
Washington, DC 20007

Canadian Lakes Loon Survey
Long Point Bird Observatory
PO Box 160
Port Rowan, ON N0E 1M0

Center for Environmental Education
Center for Marine Conservation
1725 De Sales St. NW, Suite 500
Washington, DC 20036
Phone: 202-429-5609

Center for Marine Conservation
1725 De Sales Street, NW
Washington, DC 20036
Phone: 202-429-5609
Fax: 202-872-0619

Chihuahuan Desert Research Institute
PO Box 905
Fort Davis, TX 79734
Phone: 915-364-2499
Fax: 915-837-8192
Internet: http://www.cdri.org

Coast Alliance
215 Pennsylvania Ave., SE, 3rd floor
Washington, DC 20003
Phone: 202-546-9554
Fax: 202-546-9609
Internet: coast@igc.apc.org

Desert Protective Council, Inc.
PO Box 3635
San Diego, CA 92163
Phone: 619-298-6526
Fax: 619-670-7127

Environmental Defense Fund
257 Park Ave. South
New York, NY 10010
Phone: 800-684-3322
Fax: 212-505-2375
Internet: http://www.edf.org

Environmental Network
4618 Henry Street
Pittsburgh, PA 15213
Internet: http://www.envirolink.org

Environmental Protection Agency
401 M Street, SW
Washington, DC 20460
Phone: 202-260-2090
Internet: http://www.epa.gov

Forest Watch
The Wilderness Society
900 17th st. NW
Washington, DC 20006
Phone: 202-833-2300
Fax: 202-429-3958
Internet: http://www.wilderness.org

Freshwater Foundation
2500 Shadywood Rd.
Excelsior, MN 55331
Phone: 612-471-9773
Fax: 612-471-7685

Friends of the Earth
1025 Vermont Ave. NW, Ste. 300
Washington, DC 20003
Phone: 202-783-7400
Fax: 202-783-0444

Global ReLeaf
American Forests
PO Box 2000
Washington, DC 20005
Phone: 800-368-5748
Fax: 202-955-4588
Internet: http://www.amfor.org

Global Rivers Environmental Education
Network
721 E. Huron Street
Ann Arbor, MI 48104

Greenpeace USA
1436 U Street NW
Washington, DC 20009
Phone: 202-462-1177
Fax: 202-462-4507

Internet:
http://www.greenpeaceusa.org

International Joint Commission
1250 23rd Street NW, Suite 100
Washington, DC 20440
Phone: 202-736-9000
Fax: 202-736-9015
Internet: http://www.ijc.org

International Society of Tropical
Foresters
5400 Grosvenor Lane
Bethesda, MD 20814

Rainforest Action Movement
430 E. University
Ann Arbor, MI 48109

Isaak Walton League of America
SOS Program
1401 Wilson Blvd., Level B
Arlington, VA 22209

National Project Wet
Culbertson Hall
Montana State University
Bozeman, MT 59717

National Wetlands Conservation Project
The Nature Conservancy
1800 N Kent Street, Suite 800
Arlington, VA 22209
Phone: 703-841-5300
Fax: 703-841-1283
Internet: http://www.tnc.org

Nature Conservancy
1815 North Lynn Street
Arlington, VA 22209
Phone: 703-841-5300
Fax: 703-841-1283
Internet: http://www.tnc.org

North American Lake Management
Society
PO Box 5443
Madison, WI 53705
Phone: 608-233-2836
Fax: 608-233-3186
Internet: http://www.halms.org

Project Reefkeeper
1635 W Dixie Highway, Suite 1121
Miami, FL 33160

Rainforest Alliance
650 Bleecker Street

New York, NY 10012
Phone: 800-MY-EARTH
Fax: 212-677-2187
Internet: http://www.rainforest-
alliance.org

River Network
PO Box 8787
Portland, OR 97207
Phone: 800-423-6747
Fax: 503-241-9256
Internet: http://www.rivernetwork.org/
rivernet

Sierra Club
85 2nd Street, 2nd fl.
San Francisco, CA 94105
Phone: 415-977-5500
Fax: 415-977-5799
Internet: http://www.sierraclub.org

U.S. Fish and Wildlife Service Publication Unit
1717 H. Street, NW, Room 148
Washington, DC 20240

World Meteorological Organization
PO Box 2300
41 Avenue Guiseppe-Motta
1211 Geneva 2, Switzerland
Phone: 41 22 7308411
Fax: 41 22 7342326
Internet: http://www.wmo.ch

World Wildlife Fund
1250 24th Street NW
Washington, DC 20037
Phone: 202-293-4800
Fax: 202-293-9211
Internet: http://www.wwf.org

WEBSITES

Note: Website addresses are frequently
subject to change.

Arctic Studies Center: http://www.nmnh.
si.edu/arctic

Discover Magazine: http://www.
discover.com

Journey North Project: http://www.
jnorth@learner.org

Long Term Ecological Research Network: http://www.lternet.edu

BIBLIOGRAPHY

Monterey Bay Aquarium:
http://www.mbayaq.org

Multimedia Animals Encyclopedia:
http://www.mobot.org/MBGnet/vb/
ency.htm

National Center for Atmospheric
Research: http://www.dir.ucar.edu

National Geographic Magazine:
http://www.nationalgeographic.com

National Oceanic and Atmospheric
Administration: http://www.noaa.gov

National Park Service:
http://www.nps.gov

National Science Foundation:
http://www.nsf.gov/

Nature Conservancy:
http://www.tnc.org

Scientific American Magazine:
http://www.scientificamerican.com

Ouje-Bougoumou Cree Nation:
http://www.ouje.ca

Thurston High School, "Biomes"
http://www.ths.sps.lane.edu/biomes/
index1.html

Time Magazine: http://time.com/heroes

Tornado Project: http://www.
tornadoproject.com

University of California at Berkeley:
http://www.ucmp.berkeley.edu/
glossary/gloss5/biome/index.html

World Meteorological Organizaion:
http://www.wmo.ch

INDEX

Italic type indicates volume number; **boldface** type indicates entries and their pages; (ill.) indicates illustrations.

C

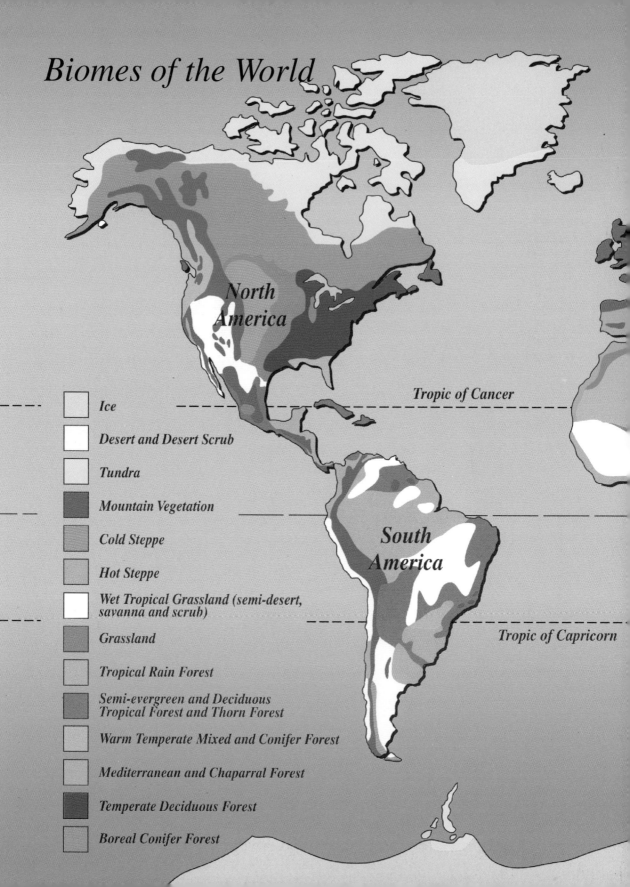

Biomes of the World

North
America

South
America

Tropic of Cancer

Tropic of Capricorn

Ice

Desert and Desert Scrub

Tundra

Mountain Vegetation

Cold Steppe

Hot Steppe

Wet Tropical Grassland (semi-desert,
savanna and scrub)

Grassland

Tropical Rain Forest

Semi-evergreen and Deciduous
Tropical Forest and Thorn Forest

Warm Temperate Mixed and Conifer Forest

Mediterranean and Chaparral Forest

Temperate Deciduous Forest

Boreal Conifer Forest